Social Forestry and Agroforestry

Social Forestry and Agroforestry

Authors

M.P.DIVYA

K.T. PARTHIBAN

K. VANANGAMUDI

K. SRINIVASAN

M. GOVINDA RAO

SSPH

SATISH SERIAL PUBLISHING HOUSE

403, Express Tower, Commercial Complex, Azadpur, Delhi-110033 (India)

Phone : 011-27672469, 27672852, Fax : 91-11-27672046

E-mail :info@satishserial.com, hkjain1975@yahoo.com

Website : www.satishserial.com

Published by :

SATISH SERIAL PUBLISHING HOUSE
403, Express Tower, Commercial Complex, Azadpur,
Delhi-110033 (India)
Phone : 011-27672852, Fax : 91-11-27672046
E-mail : info@satishserial.com, hkjain1975@yahoo.com
Website : www.satishserial.com

Reprint 2014

ISBN 81-89304-48-8

Printed at

SALASAR IMAGING SYSTEMS, DELHI-35

Tamil Nadu Agricultural University

Prof. C.Ramasamy
Vice-Chancellor

Coimbatore-641 003
Tamil Nadu, India

Foreword

The forests in the country play a significant role in the economic prosperity and ecological stability of the nation. The biological diversity in the forests contains not only the wood but also a variety of other species. There has been considerable international attention in the recent years on the issue of deforestation particularly in the tropics. Till the recent past, the deforestation rate in the country was 1.5 million ha year^{-1} which has drastically come down after the promulgation of Forest Conservation Act, 1980. This Act has facilitated the closure of natural forests for felling which created a widening gap between demand and supply of major forest products. Hence, the wood based industries in the country have felt with shortfall in raw materials supply. This created the promotion of forestry outside the traditional forest ecosystem especially in the farmlands of the country. The National Forest Policy,1988 and the National Agricultural Policy,2000 have also promoted the social, farm and agroforestry plantations through people's participation. This besides, the paucity of labour and spiraling cost of inputs, many farmers in the country also started practising tree based farming systems.

Against this backdrop, both government as well as non-governmental organizations coupled with farmers have started social and agroforestry programmes in the wastelands and marginal farmlands. Many of these social and agroforestry programmes aimed with afforestation of large stretches of wastelands resulted in variability in success rates. This might plausibly due to lack of suitable silvicultural technologies coupled with precision management systems. However, in the recent past, precision knowledge and technology have been developed towards social forestry in general and agroforestry in particular. But this knowledge and technological inputs are not available in the form of ready reckoner for use by all levels of stakeholders. In this context, the current textbook on Social forestry and Agroforestry is a significant attempt which encompasses the basic concepts of social and agroforestry coupled with suitable agroforestry models for arid, semi-arid and other agro ecosystems. Hence, this book will act as an excellent feed stock for the farm graduates as a source of reference material to learn the state-of-the-art social and agroforestry systems. This besides, the book will be useful to professional foresters and potential tree growers.

I complement the authors for their commendable efforts in bringing out the current publication for use by various levels of stakeholders.

I also wish the authors every success in all their future endeavours.

Dr. C. RAMASAMY
Vice Chancellor

Preface

Realizing the pivotal role the forests play in the economic prosperity and ecological stability, both government as well as many non-governmental organizations promote forestry activities not only in the traditional forest areas but also outside the forests in the form of social and agroforestry. These activities warrant a readymade reference material incorporating all technological inputs for use by various levels of stakeholders for implementation of successful forestry projects. After the National Forest Policy,1988, there was a paradigm shift in forest management strategies from traditional management to participatory management system. This policy also laid emphasis on promotion of social and agroforestry programmes in order to generate raw materials for wood based industries. These activities claim the need for reference material to adopt suitable tree based farming systems. Against this backdrop, the authors made an attempt to inculcate the needs of various user agencies in the form of a textbook entitled "Social forestry and Agroforestry". This textbook incorporated 15 chapters which are broadly classified in the form of fundamentals, management, applied and technological inputs.

The first three chapters dealt with the role of forests both tangible and intangible benefits along with historical evidences of various forestry activities. This besides, the role of various National Forest Policies in guiding the forest management system have also been dealt. This textbook also covered the basic aspects of scial forestry along with social forestry plantations and social forestry projects and JFM in Tamil Nadu. This book also included Design and Diagnosis of agroforestry system housed with suitable agroforestry models for different agro ecosystems

The trees play major role in augmenting edapho-climatic factors and hence a separate chapter on role of trees on soil and water conservation and soil fertility was included for use by watershed managers. Agroforestry plays a significant role in wasteland development. Hence a separate chapter on wasteland development was included which incorporated the wastelands classification along with afforestation activities through agroforestry.

The economic viability of any farming system is a matter of concern. Hence, a separate chapter on economic analysis of agroforestry systems has also been included with all tangible and intangible evaluation.

Lastly, the book inculcated the silvicultural technologies of important tree species suitable for timber, industrial wood, biofuel and energy. This book is a ready made reference material, incorporated all need based forestry technologies which will lead to promotion of social and agroforestry programmes. Initially, this book was primarily designed for farm graduates but later on tailored to suit the needs of various stakeholders who are venturing for promotion of social and agroforestry programmes.

This book is a maiden attempt and shortcomings if any, may be included in the further editions. Any suggestion to improve the contents of the book will be appreciated.

M.P.Divya
K.T. Parthiban
K. Vanangamudi
K. Srinivasan
M. Govinda Rao

Acknowledgement

The authors put forth their sincere thanks and gratitude to our respected and beloved Vice Chancellor for massive support to write the book on "Social forestry and Agroforestry" and also for his sustained and constant encouragement, meticulous guidance and critical suggestions in improving the contents and quality of the textbook.

The authors wholeheartedly express their sincere thanks to Dr.K.Vanangamudi, Dean (Agriculture), Agricultural College and Research Institute, Coimbatore for his painstaking and sustained effort in generating grant from Indian Council of Agricultural Research, New Delhi and also for his most valuable and untiring advice, lofty inspirations and critical comments for designing the contents of the textbook.

The authors sincerely express their gratitude and heartfelt thanks to the Deputy Director General (Education),Indian Council of Agricultural Research, New Delhi for munificent financial assistance towards writing this book.

The authors extend their sincere thanks and gratitude to Dr.M.Govinda Rao, I.F.S., Dean, Forest College and Research Institute, Mettupalayam for his constant support, encouragement and technological inputs rendered towards compilation of this text book.

The authors express their sincere thanks to the authors of various books, journals and periodicals which have been used as reference material in this book.

Thanks are also due to all faculty members and research scholars of Forest College and Research Institute, Mettupalayam for their co-operation and help rendered during the process of this endeavour. Special thanks are due to Dr. Era Pavendan,

Assistant Professor (Tamil), Mr. T. Mohanraj, Research scholar, Mr. V. Jamaludheen, Assistant Professor (For) and A. Maheswaran, IV B.Sc (For) for their help extended for writing this book.

Last but not the least, the authors owe lot to the Almighty for all the blessings endowed upon and for having showered the enlightened path throughout this endeavour.

Authors

Contents

CHAPTER 1

Role of Forests

Forests serve two major roles *viz.*, productive role and protective role. The productive role includes timber, fuel, non-timber forests products, etc. The protective role is that they protect and form natural resources. Through the process of photosynthesis, forests renew the oxygen supply in the atmosphere by absorbing atmospheric CO_2 and moderating the greenhouse effect. Forests provide an environment for many species of plants and animals, thus protecting and sustaining the diversity of nature. Forests clean the environment by muffling noises, buffering strong winds and stopping dust and gases. They regulate surface water runoff, moderate high and low temperatures and prevent soil erosion. In performing these functions, forests stabilize the climate and shape the landscape. Forests are also popular areas for relaxation and recreation.

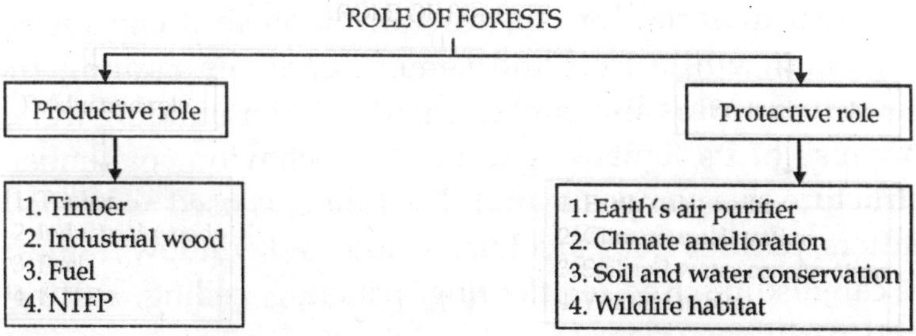

1.1. Productive role

It is estimated that the forest products contribute about 1% of world gross domestic product (GDP). The annual turnover of timber and other wood products from forests is valued at more than US$200 billion. The demand for commercial timber and other products is ever increasing, and expected to rise by 50% by 2010. Apart from that, non-timber forest products like rubber, cotton, medicinal products and so on contribute significant economic value.

1.1.1 Timber

Timber is a major forest produce and is used extensively for various purposes. In India,most of the wood produced is used for construction of houses, agricultural implements, bridges, sleepers etc., In India 12 million m^3 of timber is produced from our forests. More than 1500 species of trees are commercially exploited for timber in different parts of India. It is used in timber-based industries such as plywood, saw milling, pulp and paper and particle boards. Timber as a structural material is taking an honoured place among other structural materials because of its manifold advantages. Timber is as strong as iron and 5 to 6 times stronger than cement concrete. It is superior to iron and cement in thermal insulation, sound absorption, electrical resistance, etc. It also has a high salvage value as compared to the above materials. Timber has several good mechanical properties as well.In sea water structures timber perhaps is an ideal material for construction because it can resist corrosion while iron and cement concrete cannot. In earthquake areas also, timber is preferred for building work because of its light weight and the behaviour of timber structure as a composite unit. The timber is used as beams, rafters, purlins, trusses, columns, door and window frames. It can also be used for flooring, paneling, ceiling, roofing and partitions (Plate 1).

Plate 1. Timber and Timber Products

The major timber species are *Albizia lebbeck, Adina cordifolia,Albizia odoratissima, Albizia procera, Anthocephalus cadamba, Azadirachta indica, Artocarpus* sp., *Cedrus deodara, Dalbergia latifolia, Dalbergia sissoo, Gmelina arborea, Grewia tiliaefolia, Hardwickia binata, Hopea parviflora, Lagerstroemia lanceolata, Lagerstroemia parviflora, Lagerstroemia speciosa, Madhuca indica, Mangifera indica, Mesua ferrea, Michelia* sp., *Mimusops elengi, Ougeinia oojeinensis, Pterocarpus dalbergioides, Pterocarpus marsupium, Pterocarpus santalinus, Quercus* sp., *Shorea assamica, Syzygium cumini, Tectona grandis, Terminalia* sp., *Toona ciliata* and *Xylia xylocarpa.*

1.1.2. Industrial wood

Forests provide raw material to large number of industries e.g. pulp and paper, plywood and other boards, packing cases, matches, toys etc. (Plate 2).

The important suitable species are given below.

Pulp and paper	: Bamboos, Eucalyptus, Casuarina
Plywood	: Teak, Rose wood, Terminalia
Packing cases	: *Pinus* sp., Silver oak, Fir,
Matchwood	: Ailanthus, Simaruba, Bombax
Toys	: Adina, Redsanders, rose wood

1.1.3. Fuel

Wood is used as fuel for thousands of years, until the advent of coal, oil, gas, electricity, etc. Wood constitutes as chief source of fuel. Even today more than half of the total world consumption of wood is for fuelwood. Wood remains the major source of domestic fuel in India. Approximately 175 million m^3 of wood is used as fuel in the country. It is estimated that by 2010, most of the 3 billion people who depend on it for their daily living will find it hard to obtain. Already rural families spend precious hours in collecting

Plywood

Matchwood

Packing Case

Toys

Plate 2. Industrial wood

firewood instead of other productive work that causes losses to the tune of US$ 50 billion to the world economy (Plate 3).

The important fuel wood species are *Acacia* spp., *Casuarina equisetifolia, Prosopis, Azadirachta indica, Leucaena leucocephala,*etc.

1.1.4. NTFP

The term 'non-timber forest products' covers all forest products other than 'Major Forest Products' which consists of timber, industrial wood and fuel wood. The Non-timber forest products of commercial importance may be classified as i) Fibres, ii) Flosses, iii) Coir, iv) Grasses, Bamboos and Canes, v) Essential oils, vi) Tree Borne oil seeds, vii) Waxes, viii) Tans, ix) Dyes, x) Gums, Resins, and Oleo-Resins, xi) Drugs, xii) Spices, xiii) Edible Products and xiv) Animal, Mineral and Miscellaneous Products (Plate 4).

1.1.4.1. Fibres

The fibres of forest trees and shrubs are described under two groups viz., (i) fibres from stems and bark, and (ii) fibres from leaves.

i) Fibres from stems and bark

Fibres are obtained from the bast tissue of many woody species. Some of which yield long strong fibres, which are suitable for twisting into ropes, whereas others yield silky fibres, which are fine enough for textile purposes.

The method of extracting fibres from one another is technically known as "retting". Retting is done by soaking the fibres in water to soften the tissues. Then the fibres are beaten and scraped clean. There are certain fibres, which are used without any retting, and there are others, which are rendered useless if immersed in water.

Plate 3. Fuel wood

Plate 4. Non-Timber Forest Products

The important fibre yielding species are given below.

a) *Sterculia* spp.: *Sterculia villosa* is very common throughout India. It yields a coarse, strong, whitish pink fibre. The fibre is used extensively for making elephant harness, drag ropes and for tying rafts. The other species viz., *Sterculia urens, Sterculia foetida* and *Sterculia colorata* also yield fibres which are used for making ropes.

b) *Grewia* spp.: *Grewia tilliaefolia,Grewia laevigata* and *Grewia oppositifolia* yield coarse, strong yellow fibre used locally for rope making and domestic purposes.

c) *Hardwickia binata*: A common tree in the Peninsula.The fibre is obtained from the branches, especially from the young shoots. It is in red colour and fairly strong. It is used chiefly for cordage and rope making.

d) *Bauhinia vahlii*: This is a climber of gigantic size, commonly found in forests. It yields a valuable strong fibre used extensively for tying and other works.

e) *Acacia leucophloea*: The bark yields a fibre, which is used for fishing nets and coarse cordage.

f) *Calotropis gigantea*: This is a common shrub yielding valuable fibre. It is generally found in the stony and drier river beds. The fibre is fine, white, silky, very strong and durable under fresh and salt water. The fibre cannot be retted as it rots when so treated and it is therefore difficult to separate rapidly and cheaply. This is usually done with hand.

g) *Trema orientalis*: A small tree found in many forests of India. It yields a long light brown fibre, which is used for making ropes, twine and coarse cloth.

h) *Ficus* spp.: *Ficus religiosa* and *Ficus bengalensis* are the common trees yielding short strong fibre which is suitable for ropes.

i) *Boehmeria nivea*: It yields the well known "rhea" or "ramie" fibre which is a fine white fibre used in rope making.

ii) Fibres from leaves

The following are the some of the important species whose leaves yield fibre.

a) *Caryota urens* : This is one of the most important fibre produced from leaves. It yields the well known fibre of commerce known as kittul. This sago palm is found in the moist forests of Assam, West Bengal, Orissa, Kerala and Tamil Nadu. The fibre is of fair length and strength. The colour of the fibre is dusky brown to black.It is used by fishermen in preparing nets and fishing lines, bristles for brushes and rope making.

b) *Pandanus* sp.: The leaves of these plants produce a useful, strong fibre for cordage, fishing nets, fishing lines and sacking.

c) *Agave* spp. (*A. americana, A. cantala, A. sisalana, A. mightii*) : These are the well known aloe plants. All yield fibres of varying utility. *A. sisalana* produce the sisal fibre of commerce. This fibre is strong coarse and flexible. *A. americana* yields an excellent fibre. It is very durable for ropes and cordage.

1.1.4.2. Flosses

There are several trees and plants which produce silky flosses in their fruits. The common floss yielding species are listed below.

a) *Ceiba pentandra*: The floss from this tree is known as silk cotton or kapok. These are too short to be used for spinning or weaving, but are used extensively for stuffing purposes. The true kapok tree is found in the Western Ghats and south India.

b) *Bombax ceiba* : The floss from this tree is known as Indian Kapok. The capsule of this tree yields silky cotton which is soft and strong. The capsules should be collected when green as the floss loses much of its resilience after the capsules have opened.

c) *Cochlospermum religiosum*: This tree is known as yellow silk cotton found in many parts of the country. The floss covering the seed is resilient and can be used as a substitute for Indian or Java kapok for stuffing mattresses, pillows, cushions and for life saving belts.

d) *Calotropis gigantea*: It produces fine, silky, soft, glossy, and resilient floss, possessing good buoyancy. The floss is largely used for stuffing mattresses and pillows.

The other species which produce floss of local importance are *Calotropis procera, Holarrhena antidysenterica, Wrightia tomentosa, Cryptolepis buchanani, Salix daphnoides* and *Populus ciliata.*

1.1.4.3. Coir

The well known coir fibre used extensively all over the world for mats, brushes, ropes and other purposes is obtained from the thick husk of the fruit coconut (*Cocos nucifera*). Coir fibre is coarse, stiff, buoyant and elastic and is therefore very suitable for the manufacture of ship's ropes.

1.1.4.4. Grasses, Bamboos and Canes

i) Grasses

Grasses are used for a variety of purposes other than grazing. The grasses used for various purposes are as under.

a) *Eulaliopsis binata*: The grass is found on the bare and drier slopes of the Sub-Himalayan areas. It is common in Bihar, Orissa, West Bengal, Madhya Pradesh and Punjab. The chief use of the grass is for making paper.

b) *Saccharum munja*: The grass is found almost all over India along river beds and other places in plains and lower hills. The fibre obtained from the leaf sheath is used for cordage and ropes. The blades of the grass are also used for thatching. The lower parts of the stem are used in making chairs, stools, tables, baskets and screens.

c) *Saccharum spontaneum*: The grass is found in West Bengal and Assam and is extensively used for making ekra mud walls, which are built up with mud plastered over this grass.

d) *Vetivera zizanoides*: It produces the Khus-khus of commerce. It is found throughout the plains and lower hills upto 1500 m and along the margins of streams and lakes. Khus-khus is the name given to the aromatic and long fibrous roots, used for making the well known Khus-khus ki tatti, or mat, fans and ornamental baskets. The young grass is used for fodder and thatching purposes.

e) *Imperata cylindrica*: It is a very common grass used for thatching. Another species valued for this purpose is *Saccharum spontaneum*. The other thatching grasses are *Heteropogon contortus, Saccharum munja* and *Erianthus ravennae*.

There are many cultivated and wild grasses adapted to various climates used for feeding livestock. The more prominent grasses are *Andropogon* spp., *Apulda mutica, Bothriochloa ischaemum, Brachiaria brizantha, Cenchrus ciliaris, Chloris gayana, Chrysopogon aciculatus, Cynodon dactylon, Dicanthium annulatum, Eragrostis* spp., *Heteropogon contortus, Iseilema laxum, Panicum antidotale, P. maximum, Paspalum dilatarum, Pennisetum clandestinum, P. purpureum, and Schima nervosum.*

The following species viz., *Eulaliopsis binata, Saccharum munja, Themeda arundinacea* and *Saccharum spontaneum* are used for making rough ropes and cordage.

ii) Bamboos

Bamboos are tall, perennial, arborescent grasses, belonging to Bambusae. There are about 140 species in India. Bamboos are characterized by woody stems, commonly called culms, which arise from the rhizome. The most common, valuable and universally used of all bamboos is *Dendrocalamus strictus,* the 'male bamboo' of commerce, which is found in deciduous forests throughout the country. The culms in this bamboo are solid. *Bambusa arundinacea* is the next important one found mainly in Orissa, Assam, Southern and Western India.

The bamboos are used for a variety of purposes ranging from house posts to fountain pens. It is commonly used for roofing, rafters, walling, flooring, matting, spear and lace shafts, lathis, masts, spars, tent poles, furniture, water pipes, cart shafts, basket making, musical instruments, bows and arrows. Bamboo leaves are much valued as fodder. Bamboo seeds are eaten. The rhizomes and young shoots are pickled and eaten. Bamboo is also an important raw material for the paper industry in the country.

The important bamboos are as under.

a) *Arundinaria falcata*: It grows gregariously forming undergrowth in forests of oak and deodar, usually in moist shady places at an altitude between 1200m-2000m elevation. The culms are used for basket making, fishing rods and hookah pipes.

b) *Arundinaria spathiflora*: A small bamboo commonly found in Western Himalayas at 2000 m to 2700 m elevation as an undergrowth of fir, spruce and deodar forests.

c) *Bambusa arundinacea*: A thorny bamboo, having culms 24-30 m in height and 15-18 cm in diameter. It occurs wild, especially in the hill forests of western and southern India, ascending up to 900 m. This bamboo is used in paper manufacture.

d) *Dendrocalamus giganteus* : This is the biggest of the Indian bamboos cultivated in parts of West Bengal, Assam and Malabar. It grown upto 24-30 m high and 20-25 cm in diameter.

e) *Dendrocalamus strictus*: It occurs throughout India in deciduous forests up to 1000 m elevation. The culms are usually quite solid, 6-15 m high and 2.5-7.5 cm in diameter. It is most universally used of all the bamboos.

f) *Dendrocalamus hamiltonii:* It occurs in the sub-Himalayan tract up to 900 m. The culms are up to 25 m long, 10-18 cm in diameter and thin walled.

g) *Dendrocalamus longispathus*: This occurs in West Bengal and Assam, usually along streams up to 1200 m elevation. The culms are from 12-18 m in height and 7.5-10 cm in diameter.

h) *Melocanna bambusoides*: This bamboo is found in Garo, Khasi and Lushai hills in Assam.The culms are up to 2m in length and 9.5 cm in diameter. It is typically gregarious.

i) *Ochlandra travancorica*: This is found in the mountains of Kerala, Karnataka and Tamil Nadu. It is an erect reed like gregarious bamboo which covers large areas.

j) *Oxytenanthera nigrociliata* : This is a tufted bamboo with culms 9-15 m high and 1.3-10 cm in diameter, found in Assam, Orissa and Andaman Islands.

The other important bamboo species are *Bambusa balcooa, Bambusa tulda, Cephalostachyum pergracile, Gigantochloa macrostachya, Ochlandra scriptoria, Oxytenanthera ritcheyi* and *Pseudostachyum polymorphum*

iii) Canes

Canes, 'rattans' of commerce are the stems of climbing palms of the genus Calamus and of a few other related genera.

The genus *Calamus* consists of about 30 species occurring in India, distributed chiefly in Himalayas, Assam, Kerala, Karnataka, Tamil Nadu and the Andaman. The stems of these species are long, cylindrical and of uniform thickness, solid, straw yellow in colour. They are strong, tough and elastic. The outer surface is hard, smooth and shining.The core is spongy. The canes reach maturity in about 5 years.

Canes have a variety of uses due to their pliability, strength and long lengths. They are used as substitutes for ropes and cables, furniture frames, walking sticks, polo-sticks, umbrella handles, baskets, etc. Canes when split, the strips from the outside are used for making seats, backs of chairs and sofas. Canes are also used in wicker work and the manufacture of sports goods. The shavings from canes are useful for stuffing, packing, rough cordage and matting.

The important cane species are as follows:

a) *Calamus acanthospathus:* This is a common climber occurring in eastern Nepal, Sikkim, Bhutan, and Assam. The internodes are 15 cm long with a diameter of 2.0-2.5 cm. It yields strong canes.

b) *Calamus guruba*: This is a slender cane, 0.6-0.8 cm in diameter found in West Bengal and Assam.

c) *Calamus tenuis*: This is the common cane of North India, found in the sub-Himalayan tract.It is of medium thickness, being 0.6-1.12 cm in diameter.

d) *Calamus viminalis*: This is a thin strong cane, distributed in lower West Bengal, Orissa, Andhra Pradesh and Andaman.

e) *Calamus rotang*: This is found in central and southern India and has slender but strong stems. It is split easily and is used in furniture and basket work.

The other important canes are *Calamus andamanicus, C. brandisii, C. flagellum, C. gamblei, C. gracilis, C. latifolia, C. palustris, C. rheedei, Daemonrops kurzianus* and *Pletocomia himalayana*.

1.1.4.5. Essential Oils

The products obtained from forest plants by distillation can be divided into i) Grass oils ii) Wood oils, iii) Leaf oils, iv) Root oils, v) Flower oils and vi) Miscellaneous essential oils.

i) Grass oils

There are five well known commercially recognized aromatic grass oils. These are found in the genus Cymbopogon.

a) *Palmrosa grass oil:* Palmrosa oil is the product of rosa grass (*Cymbopogon martini*), which occurs in Madhya Pradesh, Bihar, Rajasthan, U.P., Maharashtra and Andhra Pradesh. There are two varieties viz., i) sporadic variety known as '*motia*' and ii) gregarious variety called as '*sofia*'. The oil is obtained from inflorescence while small quantities of oil can also be obtained from stem.

   Palmrosa oil is pale yellow liquid with geranium odour. It is the best source of '*geraniol*'. The oil is used as a base for the manufacture of cosmetics. It is also used for flavouring tobacco. It forms an ingredient of mosquito repelling ointments. It has several applications in medicine. It blends well with soap perfume material.

b) *Lemon grass oil:* The oil is obtained from *Cymbopogon flexiosus*. The grass occurs in South India, mainly in

Kerala and to a lesser extent in Tamil Nadu and Karnataka. The quality of lemon grass oil is judged mainly by its *'citral'* content, which varies from 41% to 85%. The oil is reddish yellow to reddish brown liquid with intense lemon odour. It is used chiefly in manufacture of scents, in perfuming soaps, for cooking and flavouring, and as an anti mosquito agent and in many preparations of pharmaceutical nature.

c) *Ginger grass oil:* Ginger grass oil is obtained from *Cymbopogon martini* var. *sofia.* It is found gregariously in parts of Maharashtra. Ginger grass oil is sweet smelling oil. It is used in perfume blends and for scenting the cheaper varieties of soap.

d) *Citronella oil:* Citronella oil is obtained from Citronella grass called *Cymbopogon nardus.* The grass grows naturally in hills of Tamil Nadu, Andhra Pradesh and north-east corner of Assam. The oil is used extensively in low cost perfumes, soaps, floor waxes, insecticides and other household products.

e) *Khus oil (Vetiver oil):* The oil is obtained from *Vetivera zizanoides,* a densely tufted grass, which is found throughout the plains and lower hills up to an altitude of 1200 m. Khus oil is used in perfumes, cosmetics and soaps. It is also used for flavouring beverages.

ii) Wood oils

The important wood oils are (a) sandalwood oil from *Santalum album,* (b) agar wood oil from *Aquilaria agallocha,* (c) deodar oil from *Cedrus deodara* and (d) pine oil from *Pinus* species.

a) *Sandal wood oil:* The sandalwood oil is obtained from the heartwood of *Santalum album,* an evergreen parasitic tree of Tamil Nadu, Karnataka, Andra Pradesh and Maharashtra. The oil has a sweet odour. The chief constituent is 'santalol', which is a perfume material and is a blender fixative.

b) *Agar oil:* Agar oil is distilled from the resinous portion of the wood of *Aquilaria agallocha.* The tree occurs in the Khasi Hills of Assam and parts of West Bengal. The oil is used in perfumery, as a perfume retainer and for blending in high grade perfumes and for burning incense.

c) *Deodar wood oil:* An essential oil known as deodar oil is obtained by steam distillation of the chips of sawdust of *Cedrus deodara.* The oil is reddish brown to golden yellow, possessing a characteristic odour. The oil is used in the perfumery, soap making industries and to adulterate sandalwood oil.

d) *Pine oil:* Pine oil is the product of the heartwood and stumps of pine trees. Pine oil is used as solvent for resins, oils, cellulose, ethers, germicides, and insecticides, etc. It is used in medicine, in veterinary disinfectant sprays, and in perfumes for soaps.

iii) Leaf oils

a) *Eucalyptus oil:* The distillation of oil is from the leaves and terminal branchlets of *Eucalyptus globulus.* The oil is colourless. It has a characteristic aromatic odour and spicy taste.The main use of the eucalyptus oil is in pharmaceutical preparations, flavouring tooth pastes, perfumery, vermin repellents, and disinfecting preparations.

b) *Citriodora oil:* The oil is distilled from the leaves of *Eucalyptus citriodora* introduced in Tamil Nadu, Maharashtra and plains of North India. The oil contains 'citronellal'. The oil is used in soap making and perfumery.

c) *Camphor oil:* Camphor and camphor oil are obtained from the wood and leaves of *Cinnamomum camphora,* grown in parts of U.P., West Bengal and South India.

Both camphor and camphor oil are used in cold remedies, perfumery and insecticides.

d) *Cinnamon oil:* The cinnamon oil distilled from leaf and bark of *Cinnamomum zeylanicum*, is used for flavouring, confectionery, pharmaceuticals, soaps and dental preparations. It has high germicidal and fungicidal properties.

e) *Mint oil (Peppermint): Mentha arvensis*, a herb found throughout the temperate Himalayas, at 1500-3000 m, growing wild, yields the mint oil. The plant is also being cultivated on a large scale. The oil contains menthol largely used in confectionery and in medicine.

iv) Root oils

The important root oils are as below.

a) *Costus oil (Saussuria lappa):* The plant is grown in Kashmir and upper Himachal Pradesh. The oil is obtained by distillation of the roots. The oil is a pale yellow viscous liquid with a peculiar, extremely tenacious odour. It is used in the blending of many types of perfumes.

b) *Valerian oil (Valeriana wallichii):* This plant is a herb growing abundantly in temperate Himalayas from Kashmir to Bhutan at an altitude of 1200-3000 m. The oil is used for flavouring and blending.

v) Flower oils

a) *Keora oil :* Of essential oils obtained from flower, the 'Keora oil' extracted from screw pine (*Pandanus tectorius*) is an important one.It is a small evergreen shrub or a small tree, found in parts of Madhya Pradesh, Orissa, Rajasthan, South India, West Bengal and Uttar Pradesh. It forms dense thickets in tidal forests. The well known perfumery products obtained from keora flowers include the Attar of keora, keora water and keora oil. The products are used in the preparation of cosmetics, perfumes and hair oils.

vi) Miscellaneous essential oils

a) *Acacia farnesiana*: The flowers yield the 'cassie' perfume of commerce.

b) *Boswellia serrata*: The gum oleo-resin yields an essential oil, which serves as a substitute for pine turpentine.

c) *Bursera delpechiana*: Linaloe oil is obtained from the heartwood and berries. It is used as a fixative in perfume and soaps.

d) *Citrus limonia*: The chief constituent of oil is citral. The oil is used in blending with other essences to give 'verbena' perfume.

e) *Jasminum grandiflorum*: The jasmine perfume is obtained from the flowers by solvent extraction.

f) *Juniperus macropoda*: Cedar oil is manufactured by distilling shavings and sawdust of the Himalayan pencil cedar.

g) *Michelia champaca*: Champ oil, which is obtained from the flowers of *Michelia champaca* is used as an 'attar' in the local perfume industry.

h) *Mimusops elengi*: The flowers on steam distillation yield an essential oil, which is used in manufacture of perfumes.

1.1.4.6. Tree Borne Oil Seeds

The following are the important tree borne oilseeds.

a) *Actinodaphne hookeri*: This is a medium tree found in the tropical evergreen forests of the Western Ghats. The fat is used for making detergents.

b) *Aleurites fordii*: The tung oil is obtained from *A. fordii*. The oil is used in paints and varnishes, manufacture of linoleum, oil-cloth, water-proofing of textiles, printing inks and electrical insulation.

c) *Azadirachta indica*: The oil is used in medicine, as an illuminant and for manufacture of soap.

d) *Calophyllum inophyllum*: The tree is found in the coastal region and locally in the hilly tracts of peninsular India. The oil is used for burning and for soap making.

e) *Madhuca butyracea*: The Indian butter tree, is found along the sub-Himalayan tract from Ganga to Bhutan. The seeds yield oil known as 'Phulwara Butter, which has the consistency of ghee. It is used as an adulterant of ghee, employed in manufacture of soaps, candles, and as an ingredient of chocolate manufacture.

f) *Garcinia indica:* The tree yields a valuable edible fat known as 'Kokam Butter'.

g) *Hydnocarpus kurzii*: Chaulmugra oil is obtained from the seeds of *Hydnocarpus kurzii*, a tree found in tropical evergreen forests throughout upper Assam. The oil is used in the treatment of leprosy and skin diseases.

h) *Madhuca indica*: The tree grows wild in forests throughout the country. The seed oil is known as 'Mahua butter'. It is used for cooking purposes, hair oil, soap making and as an illuminant. The oil finds use in medicine also.

i) *Mallotus philippensis*: The tree has wide distribution throughout the forests in the country. It's oil is used as a substitute for tung oil in making quick drying paints and varnishes, hair fixers and ointments.

j) *Melia azaderach*: It is often cultivated as an ornamental avenue tree. The seeds yield an oil which is suitable for making soaps and hair oils.

k) *Mesua ferrea*: The tree is found in abundance in Assam, West Bengal and Western Ghats. The oil is used for burning, lubricating, soap making and application on sores.

l) *Mimusops elengi*: The tree is found in the Deccan Peninsula and the Andaman Islands. It yields fatty oil which is used for edible and lighting purposes.

m) *Pongamia pinnata*: The tree is found in central and southern India. The seeds yield oil, which is used in medicines, soap making and leather tanning.

n) *Schleichera oleosa*: This tree occurs in deciduous forest throughout central and peninsular India. It produces seeds yielding the kusum oil of commerce. The oil is used for cooking, lighting and soap making.

o) *Shorea robusta*: The tree has a wide distribution in the country. The cotyledons of its seed yield the well known 'Sal Butter' used for cooking, lighting and for adulterating ghee.

p) *Vateria indica* : This tree is found in Western Ghats from Maharashtra to Kerala. The seeds yield solid oil known as 'piney tallow'. It is used for lightning, as a substitute for ghee, and for medicinal purposes.

1.1.4.7. Waxes

Waxes are usually found on the epidermis of leaves and fruits. The seeds of some trees growing in India produce wax useful for burning purposes. The important wax yielding species is *Sapium sebiferum*, which is found wild, extensively in North India. The seeds of this tree are coated with a white wax which can be used for the manufacture of candles, soaps and for dressing cloth. A another wax producing species is *Rhus succedanea*, a small tree of the Himalayas. It's seeds are mixed with the oil of *Melia azaderach* produce a wax which is made into candles and sold under the name of 'Japan wax'.

1.1.4.8. Tannins

Tannins are secretion products found in plant tissues. The name tanning is given to organic substances which have the property of combining with the albumen and gelatin to form an insoluble compound which resist decay. The tannins are found in wood, bark, leaves and fruits of many forest plants.

Tanning material can be classified into i) Bark tans, ii) Fruit tans and iii) Leaf tans.

i) Bark Tannin

There are numerous trees yielding bark useful for tanning purposes. Tan liquors are extracted from the bark by the application of steam and mixed with other extracts to give required colour and quality to leather. Some of the bark tans are described below:

a) *Acacia nilotica*: It is a common tree found in forests and wastelands. The bark is cheap and abundant. The tannin content is about 18 per cent.

b) *Cassia auriculata*: It is a small bush which grows in the south and western India. It is considered superior to babul bark and contains 23% tanning material.

c) *Cassia fistula*: A tree which has a wide distribution all over tropical India. The tannin content is 10-12%.

d) *Shorea robusta*: It's bark contains 3-9% tannin and furnishes a very tough and reddish colour to the leather.

e) *Acacia mollissima* : It is cultivated in South India. It is the most important among vegetable tan stuff. It has replaced Avaram and Konnai barks to a considerable extent.

f) *Terminalia arjuna*: It is a tree occurring throughout India usually on banks of rivers and streams, especially in central and southern India. The tannin content in bark is 20%-24%. The colour of leather produced is light brown.

g) *Ceriops roxburghiana*: This is a small evergreen mangrove tree in the coastal forests and tidal creeks.The tannin content in bark is 20-37%.

h) *Rhizophora mucronata*: *This* is another well known mangrove tree whose bark is rich in tannin, but it contains an excessive quantity of colouring matter. It is used in local tanneries.

ii) Fruit Tannin

The fruits of certain trees are used for tanning purpose and for manufacturing tannin extracts.

a) *Myrobalans:* The most important of the myrobalans is the fruit of *Terminalia chebula,* known as the chebulic myrobalan. Two other myrobalans are the beleric myrobalan from *Terminalia bellirica* and the emblic myrobalan from *Emblica officinalis.* Myrobalan tree is widely distributed and occurs in Maharashtra, Tamil Nadu, Madhya Pradesh, West Bengal, and Northern India. It's tannin material is of high quality. The average tannin content is 32%. It imparts a bright colour and is a valuable softener to the leather.

b) *Babul pods:* Another tan fruit of local value is the pod of *Acacia nilotica*. The tannin content is 12-19%. It gives a good colour to the leather and has a softening effect.

c) *Divi-divi: Caesalpinia coriaria,* which yields the divi-divi pods of commerce. The pods contain 22-40% tannin.

d) *Gothber:* The tan fruit of *Zizyphus xylopyrus* is of importance. Its fruits give black colour to the leather.

iii) Leaf Tannin

Leaves are not extensively used for tanning purposes but are used locally by the tanners as colouring agents.

a) *Anogeissus latifolia*: The tree is found in the sub-Himalayan tract. The dry mature leaves contain about 16% tannin whereas the reddish tips of young leaves are found to contain up to 55% tannin. A mixture of green leaves, red leaves and petioles, when dried and ground, yields a product containing 30% tannin. The tannin produces pale coloured leather with a greenish tinge.

b) *Carissa spinarum*: It is a thorny shrub which occurs in most parts of India. The tannin content is 9-11%, but it has the advantage that the leaves can be collected throughout the year. It is suitable for mixing with other tan material.

The other important leaf tans are *Emblica officinalis, Lawsonia inermis* and *Rhus cotinus*. Leaf galls are found on several trees and are used locally for tanning purposes eg. *Tamarix* sp, *Garuga pinnata, Prosopis spicigera* and *Terminalia* spp.

1.1.4.9 Dyes

Dyes are used in pharmacy, for dyeing leather, staining wood and, when dissolved in alcohol, dyeing cloth.These dyes can be divided into i) wood dyes, ii) bark dyes, iii) flower and fruit dyes, iv) root dyes and v) leaf dyes.

i) Wood dyes

a) *Santaline:* Santaline dye is obtained from the heartwood of *Pterocarpus santalinus*, the red sanders of South India. The wood of the tree yields a bright red dye used as a colouring agent. It gives a salmon pink colour to the cloth.

b) *Brazilin: Caesalpinia sappan* yields a red dye known as 'brazilin'. It is a tree or small shrub indigenous in the country. It is used in dyeing of cotton, silk and wool fabrics.

c) *Artocarpus:* The wood of *Artocarpus heterophyllus* and *Artocarpus lakoocha* yield a bright yellow dye used for colouring cloth. The dye is moderately fast on silk.

d) *Cutch:* The cutch extract from *Acacia catechu* is an important dyeing agent. It is used for dyeing fishing nets, sail cloth, wool and silk. It is also used for tanning and dyeing canvas, leather or cloth for marine purposes.

ii) Bark dyes

Many barks yield brown and black dyes. The following tree species yield dyes from bark.

a) *Acacia* species: The barks of *A. concinna, A. farnesiana* and *A. leucophloea* yield a black dye.

b) *Alnus* species: The dye from the bark of *Alnus spp* is used in dyeing to deepen the colour of the red and brown colour dye

c) Casuarina : The dye from the bark of *Casuarina equisetifolia* is used in dyeing fabrics and for toughening fishermen's nets.

d) Manilkara: The bark of *Manilkara littoralis* yields a red dye.

e) *Myrica esculenta*: The bark contains a yellow colouring matter, which is used in dyeing.

f) *Terminalia alata*: The bark is used for dyeing to a black colour.

iii) Flower and fruit dyes

The dyes obtained from the flowers and fruits of many forest trees are more important.

a) *Kamela:* One of the best known of the fruit dyes is obtained from the capsule of *Mallotus philippensis,* and known as kamela powder. The tree is commonly found throughout India.The dye is extensively used for dyeing silk a bright orange colour and mixed with other ingredients, the colour can be toned from pale yellow to dark red.

b) *Annatto*: The annatto dye is obtained from the seeds of *Bixa orellana,* a cultivated tree in South India. The scarlet arils surrounding the seed yield a bright yellow dye used for colouring wool, silk and calico. It is also used for colouring butter and other food stuffs. Annatto is marketed in the form of cakes, tablets or paste. The principal colouring matter of the seeds is bixin which is present to the extent of 10-12%.

c) *Dhak*: A popular flower dye obtained from the well known dhak or palas (*Butea monosperma*) tree. The flowers yield a bright yellow dye.

d) *Toona ciliata*: The flowers contain a red colouring matter, nyctanthin. It is used to dye cotton and woolen fabrics.

e) *Nyctanthes arboristis*: The orange coloured tube of the flower is rich in colouring matter, which is readily soluble in water, imparting beautiful orange or golden colour. Liquors are coloured with this dye.

f) *Mammea longifolia*: The flower buds are used for dyeing silk.

g) *Wrightia tinctoria*: An indigo like blue colouring substance is obtained from the seeds.

iv) Root Dyes

There are numerous forest trees roots yield dyes of varying importance. Some of them are given below.

a) *Berberis aristata*: Roots of this species yield a dye which is considered one of the best yellow dyes.

b) *Datisca cannabina*: The roots of the plant are being used throughout Himalayas as a yellow dye stuff for wool and cotton.

c) *Morinda careia*: The root bark contains morindone, which yields red colouring matter used for dyeing cloth.

d) *Punica granatum*: Dyes of yellow and red colour are obtained from the roots of this species.

e) *Rubia cordifolia*: A valuable red dye is obtained from the roots and stems of this plant which is common all over the Himalayas.

v) Leaf dyes

The leaves of several species of *Indigofera* contain a soluble colourless glycoside, known as 'indican', which oxidizes in water to form the insoluble indigo.

Some other forest plants also yield leaf dyes e.g. *Lawsonia inermis*. This yields the orange dye known as 'henna dye'. The leaves are dried and ground into a paste. The dye is used for fabrics, dyeing hair, eye brows, palms, fingers, toes and for other forms of personal adornment.

1.1.4.10. Gums, Resins, Oleoresins and Gum Resins

Gums and resins are exuded by plants, partly as a normal phenomenon and partly as the result of disease or injury to the bark or wood. These arise chiefly from the stem but also from the roots, leaves and other parts of the plants. "A gum is a substance of more or less sticky nature.

The resin neither dissolves nor softens in water but it is more or less soluble in organic solvents. Gum resins contain some resin and do not dissolve in water completely. Resins often occur mixed with a high percentage of essential oils and are known as oleo-resins.

i) Gums

a) *Acacia gums:* The acacia gums, largely known as 'gum arabic' are derived from several species of the genus Acacia. The following are important species of Acacia yielding gum.

1) *Acacia catechu:* The gum from this tree is darker in colour and is regarded as a better substitute for true gum Arabic and babul gum.

2) *Acacia nilotica*: Indian gum arabic or babul gum and is of great commercial importance.The gum is used in calico-printing, dyeing and as a sizing material for silk and cotton and in the manufacture of paper. This is also used in preparing sweet meats.

3) *Acacia senegal*: The true gum arabic of commerce is obtained from this species. It is a small thorny tree found in Punjab, Uttar Pradesh and parts of Rajasthan. The production of true gum arabic is negligible. The gum is usually mixed with other Acacia gums. It is used in medicine as a demulcent, in textile, mucilage, paste, polish and confectionery.

4) *Acacia modesta*: This tree yields a fair quantity of gum which is highly valued in indigenous medicine and also in calico-printing.

b) *Gum kino:* The gum is produced by *Pterocarpus marsupium*, a large tree of Central and Southern India.

The gum is of ruby red colour and contains about 75% tannic acid. It is used as an astringent in medicine.

c) *Bengal kino:* Bengal kino is the product of *Butea monosperma*. It resembles true gum kino and is used for similar purposes.

d) *Jhingan gum:* The gum is obtained from *Lannea coromandelica*. It is used as mucilage, for making ink and confectionery.

e) *Katira gum:* The term 'gum-tragacanth' is used in commerce to denote a gum which is not soluble in water but which will absorb water and swell up into a mucilaginous mass. The gum-tragacanth is not produced by any Indian tree but *Sterculia urens, Cochlospermum religiosum* and *Bombax ceiba* produce gums slightly inferior to the true gum tragacanth and called under the name of gum katira. The gums are all very astringent and are used in medicine in case of diarrhea and dysentery, as an ingredient for face creams, cosmetics and for thickening ice creams.

f) *Moringa gum:* Moringa gum is obtained from *Moringa pterygosperma*. The gum is used locally in medicine.

g) *Salai gum:* Salai gum is the product of *Boswellia serrata*. It is a gum oleo-resin of greenish yellow colour and has a pleasant aromatic smell when burnt.

h) *Dhaura gum:* The gum is derived from *Anogeissus latifolia*. The gum is extensively used for sizing paper and in calico-printing.

i) *Semla gum:* This is the gum of *Bauhinia retusa*. It is a clear gum resembling gum arabic. It is used for sizing cloth and paper and for water proofing terraced roofs.

j) *Miscellaneous gums:* The other gum yielding trees are *Albizia chinensis, A. odoratissima, A. procera, Anacardium occidentale, Azadirachta indica, Bauhinia purpurea, B. racemosa, B. variegata, Chloroxylon swietenia, Ferronia limonia, Mangifera indica, Terminalia alata* and *T. bellirica.*

ii) Resins

Resins occur in special cavities or passages in a wide variety of plants. It represents exudation products of various essential oils. They normally ooze out through the bark, hardening on exposure. Commercial resins are collected from artificial wounds. Unlike gums, resins are insoluble in water but are more or less soluble in ordinary solvents such as ether, alcohol and turpentine. They are brittle, amorphous, and more or less transparent. They possess luster and burn with a smoky flame when ignited. The resins are widely used in the manufacture of paints and varnishes, water proofing substances and soaps. They are used in medicine, pharmacy, for sizing paper, for incense, in preparation of sealing waxes and many other products.

The important resin yielding species are as given below.

a) *Pinus roxburghii*: It is the only species regularly tapped. Chir pine occurs in the Himalayan sub-tropical pine forests at an elevation between 450-2300 m, extending from Bhutan in the east to Kashmir in the west. The tree occurs in abundance and yields the resin copiously.

b) *Pinus wallichiana* : It occurs in western Himalayas in moist and dry temperate forest at an elevation between 2000-3000 m. It produces resin in lesser quantity, but the quality of resin is superior.

c) *Pinus kesiya* : The khasi pine is known to yield good quality resin.

d) *Pinus gerardiana*: This pine occurs at high altitude.

e) *Canarium strictum*: The tree is found in the forests of West Coast and Southern India. It yields the resin known as 'black dammar'. It is mainly used in varnishes, bottling wax, for plasters and for caulking boats.

f) *Hopea odorata*: The tree yields resin by natural exudation, which is known as 'rock dammar'. The tree is found in Andaman Islands. It is used for the manufacture of varnishes, as a medicine for wounds, and sores and painting pictures.

g) *Vateria indica*: This large evergreen tree occurs in the evergreen forests of Western Ghats. It yields the resin known as 'white dammar'. The resin when soft is known as 'piney resin', 'Indian copal' or 'dhupa' but when hard it is called dammar. It is used as a varnish, as an ointment for sores, incense and for caulking boats.

h) *Shorea robusta* : The resin exuded by *Shorea robusta* is known as 'ral'. It is used as incense, a disinfectant, fumigant, for paints, and varnishes.

iii) Oleo resins

Oleo resins are the name given to resins which contain a percentage of natural essential oils. These are volatile odoriferous oils, generally liquids sometimes solids, which are found secreted in inter cellular spaces, where they combine with the resins to produce exudations known as oleo-resins.

Some of the important oleo-resins species are given below.

a) *Boswellia serrata*: The tree is found in dry deciduous forests on dry hills. The tree yields a oleo-resin which is known commercially as 'salai-gum' or 'guggal'. It is used as incense, for rheumatism and nervous diseases in medicine.

b) *Dipterocarpus turbinatus*: This is a large tree of tropical forests of Assam and Andaman islands. It is a source of gurjan oil. It is used in lithographic inks, anti-corrosive coating for iron, medicinal purposes, preservation of timber and for caulking boats.

c) *Kingiodendron pinnatum*: The tree is found in evergreen forests of Western Ghats from South Maharashtra to Kerala. The tree yields a dark reddish brown oleo-resin. It is locally used as wood varnish and for medicinal purposes.

iv) Gum resins

The trees yielding gum resins are described below.

a) *Garcinia morella* : This is a small tree found in the evergreen forests of Assam, West Bengal, and the Western Ghats. It is a source of gum-resin, 'gamboge'. It is used in varnishes, lacquer and metal work. It has medicinal use as a cathartic.

b) *Commiphora mukul:* This is a small tree or shrub occurring in the arid rocky tracts of Rajasthan, Gujarat and Karnataka. The resin has an aromatic taste and odour. It is used as incense, a fixative in perfume and in medicine.

1.1.4.11. Drugs

Drugs from plants depend for their therapeutic effect upon a variety of extractives in the plants. It consists of glucosides, alkaloids or their compounds. They are obtained from fruit, flower, leaf, stem or root into solids, liquids, infusions or dusts. The drugs are pain soothers, hypnotics and stimulants. Drugs are antiseptics; astringents which have the power to constrict or contract organic tissues, draw together cut or torn tissues and check bleeding.

The drug plants are classified into i) Root drugs, ii) Bark drugs, iii) Flower, fruit and seed drugs and iv) Leaf drugs

i) Root drugs

a) *Podophyllum hexandrum* : This is a small herbaceous plant found in forests between 2500-3500 m elevation in the temperate Himalayas, especially in Himachal Pradesh, Uttar Pradesh and Jammu-Kashmir. The roots are used for diseases of liver and as a purgative in the form of pills, in the case of chronic constipation. It is a drastic but slow acting purgative.

b) *Saussuria lappa*: The roots yield the commercial kuth. The plant is common in the hills. It is used in medicine, burning as incense, useful in cough, asthma, dyspepsia and skin diseases. It is used in perfumery and keeping insects out of clothes. An oil is also extracted which is used in blending the scents.

c) *Aconitum* sp: There are about 24 species of aconites, found scattered in Himalayas between 3000-5000 m elevations. The roots are considered astringent in medicine. It has alkaloids like atisine, heteratisine, hetesine and dry hydroatisine. It acts as astringent, stomachic, and aphrodisiac. These are a valuable tonic in diarrhea, dyspepsia, cough and asthma.

d) *Acorous calamus*: The plant is found in damp marshy places. The rhizomes are remedy for colic, dyspepsia and a pleasant adjunct to tonics.

e) *Rheum emodi* : This is an herb with stout stem found in Nepal, Sikkim and Kashmir. It is used as a purgative and astringent.

f) *Picrorhiza kurrooa* : The plant is common in the alpine Himalayas from Kashmir to Sikkim. The roots are used as a stomachic, tonic and febrifuge. The active principle is the glucoside 'picrorhizin'.

g) *Valeriana wallichii* : The roots of this plant found in the temperate Himalayas are used as a stimulant, antispasmodic, and in nervous and hysterical symptoms.

h) *Asparagus adscendens*: Its roots constitute the safed musli, a valued tonic in case of diarrhoea, dysentery and general debility.

i) *Clycyrrhiza glabra*: The root is sweet, refrigerant, tonic, aphrodisiac, demulcent and alternative. The extract and decoctions of roots are used for cough, genito-urinary diseases and bronchitis. It is a mild expectorant and demulcent.

j) *Berberis aristata, B. lycium* and *B. asiatica* : From the root and lower stem wood of the plant a brown extract is prepared which is known as rasaunt. It is used in the treatment of opthalmia, for piles, as a tonic and a laxative.

k) *Rauwolfia serpentina*: It is found in the temperate Himalayas, along the foot hills and is a pink blossomed plant. Drug is obtained from the roots of this plant.It contains an alkaloid which is very potent. It lowers the blood pressure, hypertension and has a tranquilizing effect on the mentally ill. It is used for snake bite, insomnia, insanity and even blindness.

The other important root drugs are *Abroma augusta, Colchicum luteum, Hemidesmus indicus, Sarsaparilla* spp., *Nardostachys jatamansi* and *Urginea indica*.

ii) Bark drugs

a) *Cinchona ledgeriana* and *C. hybrida: The* bark yields the drug quinine which is obtained from these two species of cinchona. The trees are raised as plantations.

b) *Holarrhena antidysenterica: This* is a small tree or a shrub common throughout tropical forests. The bark has astringent, antidysentric, stomachic and tonic properties. It is used in fever and as a stimulant.

c) *Soymida febrifuga*: The drug from the bark is used in fever.

d) *Alstonia scholaris* : A tall evergreen tree of the moister regions. Its bark used as a remedy for chronic diarrhoea, dysentery and as a substitute for quinine for treatment of malaria.

iii) Flower, fruit and seed drugs

a) *Artemisia* sp: *Artemisia brevifolia* and *A. maritime* grow wild in hills at an altitude of 1200-4200 m in Kashmir, Himachal Pradesh, extending eastwards up to Kumaon. The leaves are highly aromatic. The flowers, buds and young shoots contain santonin, which are a vermifuge and a remedy for intestinal worms. It is also used as a tonic in case of dyspepsia.

b) *Strychnos nuxvomica* : This tree is found in central, southern India, West Bengal, Orissa and parts of Uttar Pradesh. It gives the product called strychnine, which is a powerful poison and is used as a stimulant in the tonic, in the treatment of nervous disorders and paralysis.

c) *Aegle marmelos* : The fruit of the bale tree is used as a tonic and astringent in case of dysentery and diarrhoea.

d) *Ricinus communis* : This is largely cultivated as a crop. It is found as a small tree scattered throughout the sub Himalayan tracts and Tamil Nadu. The 'Castor oil' obtained from the seeds is used in medicine, burning and as a lubricant.

e) *Cassia fistula:* The pulp of the long pods of this tree is used throughout India as a purgative.

f) *Chenopodium ambrosoides*: It is a much branched plant, possessing numerous lance shaped leaves. Seeds and oil are used in medicine as anthelmintic.

g) *Plantago ovata:* An annual herb whose seed coat contains mucilage, which swells up in intestines and acts as a demulcent in the inflammatory conditions of mucous membrane of gastro intestinal tract. It is used in chronic dysentery, diarrhoea and constipation.

h) *Pyrethrum*: Floral heads contain insecticidal substances, in the form of pyrethrine, which are contact poisons to insects. It is locally used in the treatment of scabies.

i) *Caesalpinia crista*: The plant yields the fever nuts, which are used extensively in case of intermittent fever.

j) *Tamarindus indica*: It yields the laxative drug which is commonly used.

iv) Leaf drugs

a) *Ephedra gerardiana*: It is found in very dry stony places in the Northern Himalayas at an elevation of 2400m. It has an alkaloid called ephedrine, which is very useful in the treatment of asthma and hay fever.

b) *Vitex peduncularis*: The leaves yield a drug when soaked in water, which is useful for black water fever.

c) *Gaultheria fragrantissima* : The leaves yield winter green oil, used in flavouring medicines, aerated water and for rheumatism. Salicylic acid is obtained from the oil and aspirin of commerce is produced.

d) *Mentha arvensis* and *M. piperita*: The plants yield peppermint oil and menthol of medicinal value.

e) *Hyoscyamus niger*: Leaves, green stalk and flowering tops contain an alkaloid hyoscyamine and hysconine. The preparations are used for counteract the gripping action of purgatives and to relieve spasm in the urinary tract.

f) *Azadirachta indica*: The neem leaves contain a smelling compound which is objectionable to insects and they are extensively used to preserve books, paper and clothes from the ravages of moths and termites.

g) *Cannabis sativa* : The plant is the source of ganja, charas and bhang, the well known narcotics which are widely used by people addicted to these products.

h) *Atropa acuminata* : The species is found wild in Kashmir and Himachal Pradesh at 1800-3600 m. The plant contains hyoscyamine and atropine. It has narcotic, sedative, and diuretic properties

i) *Datura* spp: Datura *innoxia, D. metel* and *D. stramonium* are important drug plants found in India. These are used in the treatment of asthma.

j) *Swetia chirata* : It is a herb of temperate Himalayas grows at an elevation between 1200 m and 3000 m. The dried plant parts viz., stems, leaves, flowers and roots are esteemed as a tonic, febrifuge, laxative and anthelmintic.

1.1.4.12. Spices

Spices are generally aromatic, vegetable products, used in cooking to add aroma or pungency to food and to flavour dishes. They have strong odour and sweet or bitter taste. They are generally marketed in dry conditions and are obtained from various parts of plants.

The following spices which are found in the forest are described below.

a) *Elettaria cardamomum:* It is a perennial herb indigenous in the South and West of India. It is found growing wild in many areas. The ripe fruits are used for flavouring food and for eating with pan.

b) *Piper nigrum*: It is found in the damp evergreen forests of West Bengal, Assam and the West Coast

c) *Curcuma aromatica* : The roots of *Curcuma aromatica,* a common forest plant of West Bengal and South India are known as wild turmeric and are used as a substitute for true turmeric (*Curcuma longa*), which is used as an ingredient in curries and colouring the food stuffs.

c) *Cinnamomum zeylanicum:* It is an aromatic spice. The tree is a native of Western Ghats and adjoining hills from Maharashtra southwards up to 1800 m. The bark, either as small pieces or as powder, is used as a spice or condiment.

1.1.4.13. Edible products

Many of the forest plants yield fruits which are edible. Their number is very large, and most of these are of local importance. Some of the important edible fruits are obtained from *Mangifera indica, Artocarpus lakoocha, Syzygium cumini, Juglans regia, Spondias mangifera, Zizyphus maurutiana, Garcinia zanthochymus, Pinus gerardiana* and the wild cherries, pears and figs.

1.1.4.14. Animal, Mineral and Miscellaneous Products

i) Animal products

a) *Honey:* The two common bees are *Apis dorsata* and *Apis indica*. The honey is either collected departmentally or the forests are leased to the contractors.

b) *Silk:* There are many kinds of silk worms both wild and domesticated. The best known are reared on mulberry (*Morus indica* and *Morus alba*). Other silk worms, which are found wild in *Bombax ceiba, Syzygium cumini, Terminalia alata* and *Ficus* species.

c) *Horns and skins:* Another source of revenue is from the sale of horns and skins of wild animals. Sambhar and Cheetal horns find a ready market. Rhinoceros horns are a valuable medicine.

d) *Ivory:* Some revenue is derived from the sale of ivory from tusks of elephants found dead in the forests.

e) *Bee's dammar:* The dammar bee builds a nest of small combs in hollow walls or tree trunks. The honey excreted is used for medicinal purposes.

ii) Mineral products

The important mineral products found in the forests are building stones, road metal, mica, lime stones, gravel, slates and other similar products. The collection of mineral products from forests areas is done either by mining, quarrying or surface collection.

iii) Miscellaneous products

1) *Leaves* : The leaves are used for many purposes. A few important species are given below.

a) *Bauhinia racemosa* : The leaves are used for wrapping bidis.

b) *Bauhinia vahlii* : The leaves are used for thatching, umbrellas, cups and plates.

c) *Borassus flabellifer*: It is used for thatching, making baskets, and matting.

d) *Butea monosperma* : The leaves are used for making platters and cups.

e) *Castanopsis indica*: The leaves are used for wrapping bidis, thatching and making platters.

f) *Careya arborea*: It's leaves are used for wrapping bidis, thatching and making platters.

g) *Cordia dichotoma*: It's leaves are used for wrapping bidis, thatching and making platters.

h) *Dillenia pentagyna*: The leaves are used for wrapping bidis, thatching and making platters.

i) *Diospyros melanoxylon*: The leaves are premier source of bidi leaves. Tendu leaves for bidis from

Maharashtra, Madhya Pradesh, Andhra Pradesh, Orissa, and Uttar Pradesh constitute a major cottage industry in the country.

2) *Soap nuts:* A number of plants yield products that contain a 'saponins'. The soap nuts are obtained from *Sapindus mukorossi,* a cultivated tree in Northern India in the moister tracts along the foothills of Himalayas up to 1500 m and from *S. trifoliatus* in southern India where it is found in open dry deciduous forests. Another plant *Acacia concinna,* which is a climbing shrub, yields the *Shigakai pods,* which are very good for washing hair, silks and woolen goods. Shigakai is also good for cleaning old, tarnished brassware, and for adding gloss to silk.

3) *Marking nuts:* The pericarp of the fruit of *Semicarpus anacardium* contains a corrosive juice which is used as a marking ink, with lime as a mordant. The fruit is commonly used for marking clothes and is sold in the market as a dry black drupe.

4) *Sola pith: Aeschynomene aspera* is a small, stout herb with spongy floating stem, commonly found in waterlogged places in Assam, West Bengal and South India. The stem consists of a mass of very light soft pith which is extensively used in making sola hats, toys, artificial flowers and models.

5) *Bead seeds:* There are many kinds of seeds, which are used as beads for making necklaces. Some of these are given below.

   a) Abrus precatorius : Goldsmiths use the seeds for weighing. Seeds are used as beads for necklaces and for ornamental purposes.

   b) Adenanthera pavonina : The hard, shining, scarlet seeds which are used as beads.

c) *Coix lacryma* : This is a tall erect grass found in the plains up to 1500 m. The fruits are used for making rosaries, beads and curtains.

d) *Corypha umbraculifera* : A tall palm found in West Coast yields seeds which are hard and white. These can be used for beads buttons and other ornamental articles.

e) *Elaeocarpus sphaericus* : A tree of Bihar, West Bengal, Assam, Madhya Pradesh and Maharashtra. The drupe has a stone, which is cleaned, polished and stained. These are made as beads for rosaries and bracelets. They are much prized by brahmins and sanyasis.

1.2. Protective role

Forests serve many protective role as below.

1.2.1. Earth's air purifier

Our lungs absorb carbon dioxide from the blood and infuse it with oxygen. Green plants absorb carbon dioxide during photosynthesis and release oxygen into the atmosphere in return. That is why forests are often referred to as the Earth's lungs. The production of energy in the presence of light by chlorophyll containing plant parts for the subsistence of the organism. Carbon dioxide, the major green house gas, is a major requirement of photosynthesis.

Forests form an effective sink for the carbon dioxide produced as a result of animal respiration, burning of fossil fuels, volcanoes and other natural and human induced phenomenon.

Forests play a significant role in maintaining the CO_2 balance in the atmosphere. Without sufficient forest cover, all the CO_2 released in the atmosphere will not be utilised, resulting in a higher percent of co_2 in the atmosphere. Hence forests are earth's air purifier as it absorbs large quantity of CO_2 from the atmosphere.

1.2.2. Climate amelioration

Forests control climate by moderating the effects of the sun, rain and wind. Leaves absorb and filter the sun's radiant energy, keeping things cool in summer. Forests also preserve warmth by providing a screen from harsh wind. In addition to influencing wind speed and direction, they shield us from the downfall of rain, sleet and hail. Forests also lower the air temperature and reduce the heat intensity of the greenhouse effect by maintaining low levels of carbon dioxide. Forest increase local precipitation by about 5 to 10% due to their aerographic and microclimatic effect and create conditions favourable for the condensation of clouds. Forest reduce temperature and increase humidity. It also reduces evaporation losses.

1.2.3. Soil and water conservation

Forests maintain the productivity of the soil through the addition of large quantity of organic matter and recycling of nutrients. The leaves are used as manure.

Tree crowns reduce the violence of rain and check splash erosion. Forests increase the infiltration and water holding capacity of the soil, resulting in much lower surface runoff. This in turn results in checking of soil erosion.

Forest checks floods. Forests and trees reduce wind velocity considerably. Reduction of wind velocity causes considerable reduction in wind erosion, checks shifting of sand dunes and halts the process of desertification. Forests by reducing erosion check the siltation of irrigation resources. Rapid siltation of various reservoirs in the country is the result of deforestation in the catchment areas of these reservoirs.

1.2.4. Wildlife habitat

Forests act as a habitat for wildlife. Wild animals, both big and small are losing their homes around the country.

Even in our National Forests and Grasslands, critters like the bull trout are at risk because their habitat has been damaged or decreased. The National Forest Foundation puts habitat degradation and fragmentation on its list of top threats to our National Forests and has made habitat restoration a main priority. Wildlife relies on their habitat for food, shelter, protection from predators and safe places to reproduce. Each species of animal has its own unique set of requirements for its habitat, including minimum size, biological composition and climate, which, when not met, pose threats to that species' survival. Healthy habitats require that the ecosystems which comprise them remain healthy and resilient.

Forests are home to thousands of species of wildlife, including one third of all federally listed endangered species, such as salmon, black footed ferrets and Kirtland's Warblers. National Forest lands play an instrumental role in the recovery of other endangered species, including the bald eagle, greenback cut throat trout and gray wolf.

Questions

1. Forest products contribution to world gross domestic product (GDP) is

   a) 1% b) 2%

   c) 5% d) 10%

2. Approximate utilization of wood as fuel in India is

   a) 150 m m^3 b) 175 mm^3

   c) 200 mm^3 d) 250 mm^3

3. Fibre from the wood is obtained from

   a) Bast tissue b) Cellulose

   c) Hemicellulose d) Lignin

4. 'Kittul' fibre is obtained from

   a) *Musa* sp. b) *Pandanus* sp.

   c) *Agave* sp. d) *Caryota urens*

5. Manila hemp is obtained from

   a) *Musa* sp. b) *Pandanus* sp.

   c) *Agave* sp. d) *Caryota urens*

6. Khus-khus is the common name for

   a) *Vetivera zizanoides* b) *Eulaliopsis binata*

   c) *Saccharum munja* d) *Cenchrus ciliaris*

7. Male bamboo of commerce is

   a) *Bambusa bambos* b) *Bambusa vulgaris*

   c) *Dendrocalamus strictus* d) *Arundanaria falcata*

8. Palmrosa oil is obtained from

   a) *Cymbopogon flexiosus* b) *Cymbopogon martini*

   c) *Cymbopogon nardus* d) *Vetivera zizanoides*

9. Agar wood oil is obtained from

   a) *Aquilaria agallocha* b) *Santalum album*

   c) *Cedrus deodara* d) *Pinus* spp.

10. Wintergreen oil is obtained from

    a) *Cinnamomum zeylanicum* b) *Gaultheria fragrentissima*

    c) *Eucalyptus citriodora* d) *Eucalyptus globulus*

12. Kokam butter is obtained from

    a) *Garcinia indica* b) *Melia azadirach*

    c) *Mesua ferrea* d) *Madhuca indica*

13. 'Piney tallow' solid oil is extracted from the seed of
    a) *Shorea robusta* b) *Schleichera oleosa*
    c) *Vateria indica* d) *Pongamia pinnata*

14. Japan wax is produced from
    a) *Melia azaderach* b) *Rhus succedanea*
    c) *Sapium sebiferum* d) *Scheichera oleosa*

15. Myrobalan tans are extracted from
    a) *Terminalia chebula* b) *Terminalia bellerica*
    c) *Emblica officinalis* d) All the above

16. Cutch is extracted from
    a) *Acacia catechu* b) *Caesalpinia sappan*
    c) *Pterocarpus santalinus* d) *Terminalia alata*

17. 'Indian gum arabic' is obtained from
    a) *Acacia catechu* b) *Acacia nilotica*
    c) *Acacia senegal* d) *Acacia moderata*

18. 'Gum kino' is produced from
    a) *Acacia senegal* b) *Butea monosperma*
    c) *Pterocarpus marsupium* d) *Lannea coromandelica*

19. 'White dammer' is obtained from
    a) *Hopea odorata* b) *Shorea robusta*
    c) *Vateria indica* d) *Canarium strictum*

References

Dwivedi, A.P. 1993. A text book of silviculture. International Book Distributors, Dehra Dun, India. 505 p.

Raja, R.A. 1997. Forest wealth of India. Daya Publishing House, Delhi. 296 p.

—•—

CHAPTER 2

History of Indian Forestry

The forests of India are ancient in nature and composition. They are rich in variety of flora and a wide range of fauna, avi-fauna and insects. India possesses a distinct identity, not only because of its geography, history and culture but also because of the great diversity of its natural ecosystems. The panorama of Indian forests ranges from evergreen tropical rain forests in the Andaman and Nicobar Islands, the Western Ghats, and the north-eastern states, to dry alpine scrub in the Himalaya to the north. Between the two extremes, the country has semi-evergreen forests, deciduous forests, thorn forests, subtropical pine forests in the lower montane zone and temperate montane forests.

2.1 History

History of forests in India, as all else is related to history of civilization. The civilizations of Mohanjodaro, Harappa and Channudaro in India date back to between 4000 and 5000 B.C. Evidence of wood samples and other materials gathered from the Indus Valley civilization reveal dependence of people in that period on wood as well as their love for trees. The history of indian forestry is detailed below.

2.1.1 Gupta Period

Historical annals render an account of forestry from the time of Chandra Gupta Maurya. Reliable historical documents are accounts of the Greek Ambassador Megasthenes (321-247 B.C) and the famous Arthashastra of Kautilya which depicts various aspects of forest and wildlife management. At that time, there was a separate department of forest products headed by "*Kupyadhyaksha*".

"*Kupyadhyaksha* was an officer who managed valuable timber like Sandal wood, Bamboo and other forest products. He was also responsible for collection of good quality timber, including timber harvesting, disbursement of salary and wages to staff employed in forestry operation as also penalize persons who are unlawfully felling trees. He should, however not prosecute persons cutting timber for their bonafide agricultural implements."

Kautilya's Arthashastra dealt in detail with the classification of forests and duties of forest officers, various rules and regulations regarding preservation and exploitation of forests, etc. After the regine of Chandragupta Maurya, Emperor Ashoka adopted and improved these practices. Plantation of fruit bearing and shade trees for the benefit of travelers and common people was started in right earnest. Ashoka was follower of the creed of non-violence and he was the founder creator of '*Abhayaranya*', which are presently termed as National Parks and Sanctuaries.

During Gupta period forest management practice improved and flourished. A well known work of that time '*Shukraniti*' throws light on these improvements. Notable among which were making of forests roads, plantation of fruit bearing trees and medicinal plants in the villages, appointment of '*Gaulmikas*' for collection of forest revenue. Various other literary works including '*Neetisara of*

Kamandaka' indicate that forests were in abundance and contained multiplicity of fauna as well as flora.

2.1.2 Mughal Period

Subsequently, however, destruction and mismanagement of forests started with the invasion of India by foreign powers. The tempo of deforestation and destruction, which increased during the Mughal period, reached its peak during the British regime. During the Mughal period large chunks of forests were cleared as forests provided shelter to dacoits, anti-social elements and insurgents. Frequent change of political bodies coupled with the pressure of burgeoning population and livestock, clearance for agriculture led to large scale destruction. The only redeeming feature was that some forests were reserved for game- hunting, gardens and recreation. Canal bank plantations were also started in some areas. In the *Sunnad of Emperor Akbar* it was directed that on both sides of the canal down of Hissar, trees of both for shade and blossom be planted so as to make it like the canal under the tree in Paradise and that the sweet flavour of the rare fruits may reach the mouth of every one, and that from those luxuries a voice may go forth to travelers calling them to rest in the cities where their every want will be supplied.

2.1.3 Early British Period

The forests deteriorated considerably with the advent of the British who were not familiar with systematic forestry. They considered forest wealth of India as an inexhaustible resource. Forestry was confined to only exploitation without any regard for replanting. Large quantities of teak and other valuable timber were cut down from areas easily approachable for the use of British Admiralty's Fleet. Scented sandalwood was also felled and exported to various European countries. Felling was allowed liberally and freely for development of agricultural and local settlements. A commission was appointed in the year 1800 to enquire into the availability of teak wood in Malabar forests. Ironically Captain Watson of the police department was appointed as the first conservator of Forest in 1806. He was assigned the duties of procuring maximum possible quantity of Indian timber for ship building in Britain. This resulted in plunder rather than conservation of forests, particularly in Tranvancore and Malabar. Royalty rights were declared over teak in South and its unauthorized felling was prohibited. In 1823, this post of Conservator was abolished.

To meet the demand for large sized timber, apart from teak, sal and deodar were also felled in an unrestricted manner. Although officials were mindful of the necessity of replacing the forests on areas cut over, the knowledge of how to carry out that desire was absent. To quote Stebbing "The very extensive correspondence which passed between the three Governments of Bengal, Madras and Bombay and the East India Company Directors affords incontrovertible evidence of this intention."

The first positive step towards improvement of forests came from South India in 1842 when Conolly, the Collector of Malabar and Chathu Menon an Assistant Sub-Conservator, raised pioneer teak plantations at Nilambur. These plantations were highly successful and over the years became world famous. This afforestation activity forced the Directors of East India Company to undertake improvement of forests of Malabar by forming teak plantations.

2.1.4 Charter of the Indian Forests

In 1855, Lord Dalhousie, the Governor General, promulgated for the first time an outline for forest conservancy for the whole country by the issue of a Memorandum of the Government of India dated 3.8.1855.

In 1856 rules for conservancy of trees and brushwood in Rawalpindi districts were published followed by forests conservancy rules in Hazara in 1857. This marked the commencement of the much needed protection for the forest areas of Punjab.

As a result of Indian Mutiny of 1857, the Royal Proclamation in 1858 declared the Sovereignty of Queen Victoria as 'Empress of India'. The country began to be opened up by a network of railway lines and roads. To satisfy this growing demand sound principle of conservation were all but abandoned and forest destruction went unabated. Large stretches of forest land were cleared to lay roads, railway lines and the best timber from adjoining areas converted for manufacture of sleepers, railway coaches, as well as to meet the various types of growing demands of fast developing military centres. Devastation of such a magnitude, however, led to some concern and serious thought was given to protect and develop the forest resources of the country.

At about the same time (1858-64) exotic acacias and eucalyptus species were introduced in Nilgiris. Plantation activity thus became a recognized part of forest conservancy. Publication of a book '*Forests and Gardens of South India*' by Cleghorn, Conservator of Madras Presidency catalyzed the forest conservancy works.

2.1.5 Awakening of Organized Forest Management

In the year 1864 the Government of India took the important decision of treating forests as State property. The idea of allowing proprietary rights in forests to individuals was abandoned and all government forests were made inalienable. The need for exercising great care in disposal of waste lands containing forests, demarcating and fixing the limits of forests to be preserved as such and enacting a comprehensive Indian Forest Act carrying out an All India Forest policy were some of the important items emphasized by the government. The systematic management of forests began with the appointment of *Dietrich Brandis,* a trained German forester, as the first Inspector General of Forests in 1864. The objective of management of forests changed from obtaining supplies of timber for various purposes to protecting and improving forests and treating them as a biological growing entity.

The first Indian Forest Act was enacted in 1865 to be superceded by new Acts in 1878 and 1927. The Forest Act provided the foundation for consolidation, reservation and demarcation of forests as well as their protection. Reserve and protected forests were categorized. In the reserve forests everything was prohibited unless specifically permitted and in a protected forest everything was permitted unless specifically prohibited. In 1894 the first forest policy of the country was formulated. This policy had two major enunciations: first, the claims of cultivation were given priority over forests preservation, and secondly, the public benefit was the sole object of forest administration. The provisions of the Forest Act, to begin with, were first applied to forests, which were important from the point of view of protecting natural water supply. Sizable forests areas were left as wastelands for the use of people living in villages and their control vested with revenue authorities. The condition of these forests deteriorated with the passage of time and they became eye-sores on the landscape. Regular forestry education was started in 1881 first for the rangers and later for the officers. Management plans were prepared for most of the demarcated forests keeping in view the soil and water conservation needs and silvicultural requirements of the crops. As a result government forests generally improved in stock and quality. Other forests in the princely states, belonging to zamindars and other individuals were also well preserved. The management plans (known as Working Plans in the parlance of forests) were prepared to ensure that annual cut did not exceed the annual increment in forests.

In the year 1891 Provinical Forest Service was inaugurated to recruit forest officers in India itself. The Imperial Forest Research Institute was established in the year 1906 to organize research on various aspects of forestry. During the period 1871-1900 considerable progress was made in technical education, training, forest administration, afforestation and fire protection.

As a result of the move for the Indian Home Rule in 1916 and the Non-cooperation Movement that followed, there was a general defiance against forest laws and damage to forests was caused by fires and unnecessary fellings in Bengal, Uttar Pradesh, Orissa and Andhra Pradesh as a protest against the British Rule. In the political changes of 1921, forests came to vest in the government of the provinces concerned. By this time forest organization had developed considerably and the only provinces without a Chief Conservator of Forests were Bihar, Bengal, Orissa and Assam.

2.1.6 World Wars

The World Wars not only interrupted, but also reversed the march of the country towards scientific forestry. During this period timber, fire wood, charcoal, hay and other forest produces were needed in large quantities and at short notices for meeting various war needs. Forestry was thus geared to one single purpose, to meet the demands of fighting forces. Forests were cut recklessly and the war thus took a heavy toll of Indian forests, both government as well as private. The government forests were cut regardless of the sequence of fellings laid down in the working plans, while those under private ownership also suffered as skyrocketing prices were too tempting to be resisted. It was in between a period of about 20 years that separated the two world wars that significant headway towards scientific working of forests could be made. This included a preparation of working plans for nearly all the areas declared as reserved forests, development of natural regeneration techniques for deodar in Himalayas, sal in various parts of the country and artificial regeneration of teak in south India. During this period various techniques of soil and water conservation as well as reclamation of ravines and other eroded areas were perfected which had the far reaching repercussion on problematic 'Chos' of various areas of Punjab. During this period successful plantations were raised in some of the refractory areas also. There was considerable development in roads, buildings, division of forest areas into various management units, making up of inventories for forest working and other administrative developments. Some good concepts introduced during this period were fire protection, building up of the forests capital and methods of controlling yield. The biggest thing was that forests were no longer treated as an inexhaustible reserve for extension of cultivation, 'a hunters dream and a loggers monopoly' and came to be regarded as a biological system whose productivity could be sustained only through care and scientific management.

2.1.7 Forestry in Independent India

Almost all the British officers left the arena and Indian officers assumed charge at the centre as well as in the provinces. Further political changes led to merger of princely states to adjoining provinces and reorganization of new provinces. Another major political change was abolition of Zamindari in all states of India resulting in vesting of large areas of private forest in the government. Anticipating that the private forest would eventually slip away from their hand, many owners tried to make most from the property they were ultimately going to loose.

The above political changes coupled with reckless fellings of the war years posed onerous responsibility on foresters of independent India. Systematic management was introduced in forest areas including those which formed parts of the ex-princely States. States like Bhopal and States in Central and Eastern India trained foresters from neighbouring states. Forest areas, which were vested as a result of Zamindari Abolition Acts, were surveyed, demarcated, rights and concessions determined and the legal status was made clear. A major duty of the forest services in the early 1950's became the control of the organization and management of forests. An important feature of the post Independence period was rapid extension of agriculture at the cost of forest. The net area under agriculture which was 118 million hectare in 1850-51 rose to 143m ha. in 1973-79. It was estimated that this increase was due to the diversion of 2.2 million hectare of forest area and 22 million hectare of area being categorized as cultural waste, tree crops, groves and pastures. This had a harmful effect on the overall environment and gave rise to several problems of floods, droughts, soil erosion, etc. In 1952 the old forest policy was revised. It recognized the protective functions of forests, discarded the notion that forestry had no intrinsic right to land but may be permitted on residual land not required for any other purpose; advocated scrupulous regard for sustained yield for forest management. The forest policy stipulated that 1/3rd of the total land area of the country should be maintained under forest, the proportion being 60% in mountainous regions which are more prone to erosion. As regards to grazing it was enunciated that continuous and cheap grazing should be allowed; furthermore, grazing by goats should be excluded and grazing by sheep should be restricted.

During the years 1950-52 new efforts viz., (1) to adopt a National Festival of Tree Planting (Van Mahotsav), (2) to devise measures for wild life conservation and (3) to place soil and water conservation on an all-India footing were begun. A new feature of this period was the development of international interest in Indian forestry under the auspices of Forestry Division of F. A.O. (Food & Agricultural Organization) of the United Nations

Forestry during the various plans aspired much, but suffered badly due to financial constraints. The managers of State finances used a double-edged financial sword against forestry by seeking maximum of revenue yet minimum of expenditure on forest protection and conservation. The surpluses made by state government from forests ranged from 55.50 per cent to 126.08% of monetary investment made and forestry in India was reduced to a commercial enterprise rather than an applied science till 1981. Surprising yet true that allocations to forestry from 1st to 6th plan could not surpass the figure of 1% of the total outlay in public sector, thus leaving a yawning gap between aspirations and efforts. In spite of these due to dedicated efforts of foresters, forestry advanced ahead.

During the first two plans (1951-61) survey, demarcation and rehabilitation of degraded areas, prevention of soil erosion and halting the extension of Rajasthan desert received major priority. Plantations of quick growing species like eucalyptus, poplars to meet industrial supplies were also undertaken during this period.

During the 3rd and 4th plans special emphasis was laid on enhancement of productivity by artificial regeneration, plantations of fast growing species particularly for pulp and paper. An event of major importance was the constitution of the Indian Forest Service in 1966. During the period 1969-74 elimination of middlemen and contractors was undertaken, together with state trading in timber and other forest products was started to enhance the revenue, to minimize pilferage, to ensure fair wages to forest labourers. Concomitantly the birth of State Forest Corporations brought a new phase in Indian forestry.

Another importance event affecting forestry was the appointment of the National Commission on Agriculture in 1970. Its report, which was presented in 1976, constituted a full volume (Vol. IX) on forestry.

In its recommendations it was made clear that the country could not depend for forest produce on traditional forests alone and that extending forest activity outside the existing forest areas was imperative. It recommended that to meet the community needs of agricultural timber, fuel wood, grazing and recreation; social forestry must be practiced in all areas available for the purpose. As a result of this social forestry activities were extended all over India. Social forestry schemes were formulated and implemented through foreign aid agencies like the World Bank, Swedish International Development, Agency, United States Agency for International Development, Canadian International Development Agency, Danish International Development Authority, and Overseas Development Agency of U.K. etc. A significant step taken in social forestry was the inclusion of afforestation and social forestry including farm forestry in the 20-point economic programme of the Prime Minister, as item No. 12. Thus a hitherto neglected subject was recognized as an item of prime importance towards nation building. Institutional financing for production forestry and manmade forests was an important change as forestry in India had earlier subsisted exclusively on government financing. Another very important event was a constitutional amendment bringing forests on the intervene in forestry affairs which had so far been the monopoly of state governments.

With the beginning of the 1980's ecological perceptions and social security received priority. Development without degradation became the theme of the 6th plan. Forestry development was linked with tribal economy, more national parks and wild life sanctuaries were constituted, protection of natural forests and their regeneration also received adequate attention. A very important legislative development was the enactment of Forest Conservation Act, 1980 which prohibited de-reservation of forests and diversion of forestland to non-forest use without prior concurrence of the Central Government. As a result of this Act diversion of forestland was drastically reduced. Whereas the rate of diversion in 30 years from 1951-80 worked out at a colossal figure of over hectares per annum. In many states other legislative changes followed like the Tree Protection Act in rural areas for U.P. A separate Ministry of Environment and forests was constituted in 1984.

The 7th plan had *'Forest for Survival'* as its theme. It aimed at preserving biological diversity, increasing the vegetal cover by massive afforestation and developing forestry as a people's movement, enhancing productivity of existing forests and meeting basic needs of the people e.g. firewood, fodder, small timber, minor forest produce etc. The creation of National Wasteland Development Board in 1985 has been an important step in the direction of reclaiming wastelands and greening the continent. Integrated watershed development in the Himalayan region, practice of agro-forestry, genetic improvement, research trials on propagation of various species, development of tissue culture, methodology to minimize waste and maximize production are important issues which have received adequate attention during this period.

The Government of India revised its Forest Policy of 1952 in December 1988. Pointing out the inadequacy of protection measures, diversion of forest land to non-forest uses without ensuring compensatory afforestation and essential environmental safeguards and tendency to look upon forest as revenue earning resource; a new strategy of forest conservation has been evolved. Conservation includes preservation, maintenance, sustainable utilization, restoration and enhancement of the natural environment.

The basic objectives include efficient utilization of forest produce, maximizing substitution of wood, conservation of national heritage (flora and fauna), increasing tree cover in the country through the massive people's movement with involvement of women.

Forest policy (1988) documents also stress the following.

i) Non- regularization of encroachments.

ii) Deploying improved modern management practices to deal with forest fires

iii) Increase in forest extension activities.

iv) Emphasis on forest education and research

v) Inadequacy of data regarding forest resource, need for completing survey of forest resources, updating information with recourse to modern technology and equipment.

Some other important landmarks in Indian forestry in the post independence era are constitution of various high level advisory bodies to review and recommend forest development work. Notable amongst them are Central Board of Forestry, Central Board for Wild Life, Forestry Commission, Forest Survey of India, Indian Institute for Remote Sensing and Photo-Interpretation and Wild Life Management Institute. It is of interest to note that State Agricultural universities started College of Forestry at the undergraduate, postgraduate and doctoral levels.

2.2 Forest Types in India

Sixteen forests types are recognized and subdivided into 221 minor types. Structure, physiognomy and floristic are used as characters to define the types. The main areas of tropical forest are found in the Andaman and Nicobar Islands; the Western Ghats, which fringe the Arabian Sea coastline of peninsular India; and the greater Assam region in the north-east. Small remnants of rain forest are found in Orissa state. Semi evergreen forest is more extensive than the evergreen formation partly because evergreen forests tend to degrade to semi-evergreen with human interference. There are substantial differences in both the flora and fauna between the three major rain forest regions.

The Western Ghats Monsoon forests occur both on the western (coastal) margins of the ghats and on the eastern side where there is less rainfall. These forests contain several tree species of great commercial significance (e.g. *Dalbergia latifolia, Pterocarpus marsupium, Tectona grandis* and *Terminalia crenulata*), but they have now been cleared from many areas. In the rain forests there are enormous number of tree species. At least 60 percent of the trees of the upper canopy are of species which individually contribute not more than one percent of the total number. Clumps of bamboo occur along streams or in poorly drained hollows throughout the evergreen and semi evergreen forests of south-west India, probably in areas once cleared for shifting agriculture.

The tropical vegetation of north-east India (which includes the states of Assam, Nagaland, Manipur, Mizoram, Tripura and Meghalaya as well as the plain regions of Arunachal Pradesh) typically occurs at an elevation up to 900m. It embraces evergreen and semi evergreen rain forests, moist deciduous monsoon forests, riparian forests, swamps and grasslands. Evergreen rain forests are found in the Assam Valley, the foothills of the eastern Himalayas and the lower parts of the Naga Hills, Meghalaya, Mizoram, and Manipur where the rain fall exceeds 2300 mm per annum. In the Assam Valley the giant *Dipterocarpus macrocarpus* and *Shorea assamica* occur singly, occasionally attaining a girth up to 7 m and a height up to 50 m. The monsoon forests are mainly moist sal (*Shorea robusta*) forests, which occur widely in this region.

The Andamans and Nicobar islands have tropical evergreen rain forests and tropical semi-evergreen rainforests as well as tropical moist monsoon forests (IUCN, 1986). The dominant species is *Dipterocarpus grandiflorus* in hilly areas, while *Dipterocarpus kerrii* is dominant on some islands in the southern parts of the archipelago. The monsoon forests of the Andamans are dominated by *Pterocarpus dalbergioides* and *Terminalia spp.*

2.3 *Status of Indian Forests*

As per the State of forests report of the Forest Survey of India (2003) , the actual forest cover of India is 20.64% of the geographic area, corresponding to 63.3 million ha. Only 38 million ha of forests are well stocked (crown density above 40%). This resource has to meet the demand of a population of 950 million people and around 450 million cattle. As such, country has to meet the needs of 16% of the world's population from 1% of the world forest resources. The same forest has also to cater for the 19% of the world cattle population.

Questions

1. The state having the largest forest cover as percentage of total geographical area.

   a) Arunachal Pradesh b) Mizoram

   c) Manipur d) Madhya Pradesh

2. The major forest types in India are classified into

   a) 5 b) 16

   c) 22 d) 32

3. The present total forest cover in India is
   a) 15.70% b) 20.64 %
   c) 22.50 % d) 25.10 %

4. Evergreen forests in India is around
   a) 38.5m ha b) 78.9 m ha
   c) 64.5 m ha d) 4.5 m ha

5. Tropical evergreen rain forests is present in
   a) Tamil Nadu b) Kerala
   c) Andaman & Nicobar Islands d) None of the above

References

Dwivedi, A.P. 1993. A text book of silviculture. International Book Distributors, Dehra Dun, India. 505 p.

Dietrich Brandis. 1994. Forestry in India: Origins and early developments. Natraj Publishers, Dehra Dun. 174 p.

Negi, S.S. 1994. Indian forestry through the ages. Indus Publishing Co., New Delhi. 248 p.

—•—

CHAPTER 3

National Forest Policies of India

Forest Policy is a set of principles or guidelines, adopted by a government to attain fixed objectives in forestry for the welfare of the nation as a whole. Forest Policy has to undergo changes according to the changed circumstances. The old Forest Policy presented different broad outlines for the promotion of forestry. The latest Forest Policy has attained new dimension in India to solve problems arising out of rising trends of environmental hazards caused mainly by deforestation.

The National Forest Policies of independent India are discussed below.

3.1 National Forest Policy, 1952

In their Resolution No. 22-F, dated the 19th October, 1894, the government of India in the Department of Revenue and Agriculture enunciated in broad outlines the general policy to be followed in the management of state forests in the country. During the interval that has since elapsed, developments of far-reaching importance have taken place in the economic and political fields. The part played by forests in maintaining the physical condition in the country has come to be better understood. The country has passed through two world wars which disclosed unsuspected dependence on forests. The reconstruction schemes, such as river-valley projects, development of industries and communications, lean heavily on the produce of forests.

While the fundamental concepts underlying the existing forest policy still hold good, the government of India consider that the need has now arisen for a re-orientation of the forest policy in the light of the changes which have taken place since it was enunciated.

3.1.1 Vital National Needs

The National Forest Policy, 1952 is formulated on the basis of six paramount needs of the country, viz.,

(i) the need for evolving a system of balanced and complementary land-use, under which each type of land is allotted to that form of use under which it would produce most and deteriorate least

(ii) the need for checking (a) denudation in mountainous regions, on which depends the perennial water supply of the river system whose basins constitute the fertile core of the country, (b) the erosion progressing space along the treeless banks of the great rivers leading to ravine formation, and on vast stretches of undulating wastelands depriving the adjoining fields of their fertility and (c) the invasion of sea-sands on coastal tracts, and the shifting of sand dunes, more particularly in the Rajputana desert

(iii) the need for establishing tree lands, wherever possible, for the amelioration of physical and climatic conditions promoting the general well being of the people.

(iv) the need for ensuring progressively increasing supplies of grazing, small wood for agricultural implements and in particular of firewood to release the cattle dung for manure to step up food production.

(v) the need for sustained supply of timber and other forest produce required for defence, communications and industry

(vi) the need for the realization of the maximum annual revenue in perpetuity consistent with the fulfillment of the needs enumerated above.

These vital needs indicate the functions of forests are to fulfill, and provide the fundamental basis of the policy governing their future.

3.1.2 Functional classification of forests

Having regard to the functions afore stated, the forests of India, whether state or privately owned, may be conveniently classified as follows.

(i) Protected forest, i.e. those forests which must be preserved or created for physical and climatic consideration.

(ii) National forests, i.e. those which have to be maintained and managed to meet the needs of defence, communications, industry and other general purposes of public importance.

(iii) Village forests, i.e. those which have to be maintained to provide firewood to release cow-dung for manure, and to yield small timber for agricultural implements and other forest produce for local requirements, and to provide grazing for cattle.

(iv) Tree lands, i.e. those areas which though outside the scope of the ordinary forest management are essential for the amelioration of the physical conditions of the country.

This classification is merely illustrative and is by no means mutually exclusive. In fact every forest performs more than one function, and has, therefore, to be so managed as to achieve the highest efficiency in respect of the chief functions assigned to it. This functional classification has also no bearing on the classification of forests distinguished in the Indian Forest Act XVI of 1927, which is based on the degree of control exercisable in them.

3.1.3 Necessity of classification

This broad functional classification of forests is necessary to focus attention on the kind and object of management necessary in each case. Every sizeable forests, whatever its composition, location, or category, serves both a protective and a productive purpose, and in its utility may be of local, regional, or national significance. The fact, however, must be realized that the country as a whole has a vast stake in the conservation of all forests, irrespective of their functions and ownership, and therefore, all of them should be administered from the point of view of national well-being.

3.1.4 Two possible considerations

Two considerations, plausible, no doubt, at first sight, if given undue weight to, destructive of national well-being in the long run, should be combated. They are :

(i) Neighbouring areas are entitled to a prior claim over a forest and its produce

(ii) Agricultural requirement has a preferential claim over forest lands

a) Claims of Neighbouring communities

Village communities in the neighbourhood of a forest will naturally make greater use of its products for the satisfaction of their domestic and agricultural needs. Such use, however, should in no event be permitted at the cost of national interest. The accident of village being situated close to a forest does not prejudice the right of the country as a whole to receive the benefits of a national asset. The scientific conservation of a forest inevitably involves the regulation of rights and the restriction of the privileges of user depending upon the value and importance of the forest, however, irksome such restraint may be to the neighbouring areas. The Himalayan forests, for instance, are the greatest of national assets, to them we owe the richness of the country. The denudation and under development of the Himalayan slopes leads to greater intensity and frequency of floods, recurrent erosion, and to coarse detritus being deposited on the fertile submontane tracts. This process inflicts immeasurable loss and misery on the unsuspecting million in the Indo-Gangetic plain and brings about a progressive and permanent impaivment of of soil fertility, and a cumulative reduction in the agricultural potential of the whole land. While, therefore, the needs of the local population must be met to a reasonable extent, national interests should not be sacrificed because they are not directly discernible, nor should the rights and interests of future generations be subordinated to the improvidence of the present generation.

b) Relinquishment of forestland for agricultural purposes

The indiscriminate extension of agriculture and consequent destruction of forests have not only deprived the local population of fuel and timber, but have also stripped the land of its natural defences against dust-storms, hot desiccating winds, and erosion. The old policy, which envisaged the relinquishment, subject to certain safeguards honoured only in their breach, of even valuable forest land for permanent cultivation, has resulted in general deterioration of physical conditions to the detriment of national interests and must, therefore, be given up. In the abstract, the claims of agriculture undoubtedly appear stronger than those of forestry. The notion widely entertained that forestry, as such, as no intrinsic right to land but may be permitted on sufferance on residual land not required for any other purpose, has to be combated. The role of forests in the national economy, both protective and productive, entitles forests to lay claim to an adequate share of land. The importance of tree lands in the rural economy of this region where agriculture constitutes the main-stay of the vast bulk of population can scarcely be over-emphasized.

3.1.5 Land use

The correct solution of the land problem is to evolve a system of balanced and complementary land use, under which each type of land is allotted to that form of use under which it would produce most and deteriorate least. A detailed survey of lands with a view to their proper utilization is, therefore, highly desirable.

3.1.6 Protection forests

Protection forests denote forest found, or required on hill slopes, river banks, sea-shores, or other erodable localities. In such sites the need for forest cover is dictated by purely protective physical consideration, such as prevention of erosion, conservation of moisture and control of rushing torrents and floods. The role of such forests in saving the soil from being washed away, and when maintained in catchment areas, in the prevention of floods and maintenance of stream-flow, cannot be over-emphasised. On flat country with loose sandy soil, especially under dry conditions, forests, whether natural or artificial, perform an essential function in minimizing wind erosion, fixing the soil and preventing the formation of sand dunes, and mitigating the desiccation of agricultural crops leeward of the tree cover. The national forest policy requires therefore, an immediate and speedy programme for the reconditioning of the mountainous regions, river valleys and coastal lands by establishing protective forests over larger areas, and preserving the existing ones. The primary object of management of such forests should be to utilize in full their protective influence on the soil, the water regime, and the physical and climatic factors of the locality and the interests to be thus protected should far outweigh those which it may be necessary to restrict. The scientific management of such protection forests, wherever possible, should include the production and exploitation of timber within the limits of safety.

3.1.7 Reconditioning of hills and dales

The progressive denudation of hill sides with serious repercussions on the fertility of the land, and the growing erosion along the banks of rivers, of which the Yamuna, the Chambal, the Mahi, the Narmada, the Kosi and the Damodar are notorious examples, constitute the major considerations demanding immediate attention.

3.1.8 The immobilization of the desert of Rajputana

Attention also needs to be drawn here to the Rajputana desert, more particularly to the fixation of the shifting sand dunes. Strong winds that develop in this region during the summer, transport vast quantities of sand and salt from the sea and Runn of Cutch, whipping the desert into terrific dust storms, the fury of which is felt throughout the north-western India. The desert has spread through the ages causing the westering of the Indus and the northerning of the Sutlej meeting an obstruction of sort only along its eastern confines in the Aravallis. The immobilization of the desert and protection of the remaining fertile belts inside it constitutes one of the planks of the National Forest Policy.

3.1.9 National forests

National forests constitute the basis of India's strength and wealth, for they comprise valuable timber bearing regions the produce of which is indispensable for defence, communications and vital industries. They have to be managed chiefly in the interest of the nation as a whole, and their organization and development is one of the most important functions of the states. Their management on scientific and business lines is essential for maintaining a sustained supply of wood for industry and of large timbers for defence, communications and other national purposes. The basic policy, so far as such forests are concerned, must be to attain national self-sufficiency in these vital supplies. Future development should, therefore, be directed to that end. Cultivation should not be permitted to encroach upon these valuable timber bearing tracts. The solution of the food problem of an ever-increasing population must be sought primarily in intensive cultivation and not in weakening the very basis of national existence by encroaching upon such forests.

3.1.10 Village forests

Village forests popularly termed fuel forests are intended, in the main, to serve the needs of the surrounding villages in respect of small timber for housing and agricultural implements, firewood, leaves for manure and fodder, fencing thorns, grazing and edible forest products. The supply for such requirements should be made available at non-competitive rates, provided they are utilized by the villagers themselves and not traded in. The management of such village forests should aim at meeting the present as well as the future needs of the local population. Removal of the produce in excess of its annual growth should not, therefore, be permitted. Restrictions should be imposed in the interests not only of the existing generation but also of posterity. These considerations render the entrusting of the management of village forests to panchayats, without appropriate safeguards, a hazardous undertaking as has been demonstrated in some of the states. The cooperation of panchayats should be enlisted in the protection and creation of village forests, and in the distribution of forest produce assigned to meet the needs of the local population, but not at the cost of economy and efficiency. While the profit motive in the management of these forests should be relegated to the background there is no justification for allowing them to become a burden on the general tax-payer: the expenses for development and maintenance of such forests must come from their own income.

3.1.11 Tree lands

Although tree lands are not part of regular forests, in a country like India where their increase, management and development are vital to the needs of the people, they cannot well be left out of any well considered policy. The land transformation programme of the government of India envisages the planting of 30 crores of trees in ten years, but this number is very far from about 2,000 crores of trees, which would be necessary to restore the hydrological nutritional balance of the country. The creation of forests by state forest departments on such an elaborate scale is ruled out at present by lack of funds and trained personnel. The only way in which some progress can be achieved is by making the whole national tree conscious. Such consciousness will stimulate private efforts at tree planting as has been demonstrated by the success of the National Van Mahotsava movement. It will also arrest the vandalism which feels no scruples in cutting down valuable trees, and create among the populace an urge to secure the protection of trees-a virtue as much to be desired as it is rare.

3.1.12 Scope for increasing tree lands

State government has a vast scope for an all around increase in the area under treelands. Defence, railways, public works departments, universities and colleges, boards, municipalities and other local authorities, association and institutions can lend a helping hand by converting the lands at their disposal into treelands. The new forests policy, therefore, envisages a concerted and supreme effort on the part of various governments and other agencies towards planned afforestation with a view to the enlargement of treelands. The exploration of the possibilities of such a development by the central/state is clearly indicated. A systematic programme of extending existing treelands and establishing new ones should be framed by the governments concerned. Under the new policy, it should be the duty of the forest departments concerned.

(i) to awaken the interest of the authorities within their region in the development extension, and establishment of tree lands,

(ii) to draw up plans for such purposes bearing in mind in the need for species of commercial importance,

(iii) to establish nurseries and seed stores in each area for the supply of saplings, plants and seeds,

(iv) to supervise the planting of trees and render such technical assistance as may be necessary for the development of tree lands, and

(v) to arouse tree consciousness among the people by publicity, by celebrating the Vana Mahotsava and by encouraging the Vana Premi Sangh.

3.1.13 Tree lands in agricultural areas

The importance of tree lands in the rural economy of the regions where agriculture constitutes the main stay of the vast bulk of the population cannot be over emphasized. Experience gained during the first two Van Mahotsava indicated a very considerable response in the countryside, where government officers had prepared the ground and created the necessary enthusiasm among the people. A campaign inducing villagers to plant trees in village commons and along roadsides, on the condition that they would enjoy the benefit of the fruits, timber and other produce of trees planted by them has yielded excellent results and is well worth an extended trial. The essence of success in such ventures lies in invoking the willing cooperation of the local villagers, the necessary technical guidance and help being furnished by the forest and other departments. In most localities, a cultivator has no land to utilize for raising trees, there is, however, nothing to prevent him from growing at least a few trees per acre on his own field. Much useful work in this direction has been done in the western district of Uttar Pradesh where cultivators have raised a fair amount of babul (*Acacia nilotica*) in their fields. Other species may prove to be of equal utility in other regions.

3.1.14 Control of private forests

The ownership of private forests in states where they still exist vests in individuals. Such ownership must, however, be regulated in the national interests so that the indiscriminate exercise of individual rights may not prejudice, or endanger, general welfare Regulation and control of private forests by the state on physical, climatic and economic grounds is therefore, imperative. Recent legislation in various states has assumed the following pattern

(i) Owners of private forests should in the first instance be given an opportunity to manage the forests in accordance with an approved working plan.

(ii) In the case of only recalcitrant owners who are tempted to sacrifice their capital for immediate gain, should the management of their forests be made to vest in government by the process of law.

The ownership of such vested forests should remain, however, unaffected, and the transfer should relate only to management, the net profits arising therefrom, if any, accruing the owner.

The object of the legislative measures outlined above stand in grave risk of being defeated by the tendency discernible among owners of private forest to cash in their assets by excessive exploitation of forests for personal ends. In order to arrest such destruction of forests, the National Forest Policy requires that their control and management should be strictly regulated, and where that cannot be done, they should be taken over by the State Governments by effective legislation.

3.1.15 Proportion of forest areas

The proportion of land to be kept permanently under forests would naturally vary in different regions. Practical consideration suggests, however, that India, as a whole, should aim at maintaining one-third of its total land area under forests. As an insurance against denudation a much larger percentage of the land, about 60 per cent should be kept under forests for their protective functions in the Himalayas, the Deccan, and other mountainous tracts liable to erosion. In the plains, the ground is flat and the erosion is normally not a serious factor, the proportion to be attained should be placed at 20 per cent, and in view of the pressure of agriculture effort at the extension of treelands should be concentrated on river banks and other convenient places not suitable for agriculture. At the same time it must be realized that even distribution of forests in all physical regions is as important as its overall proportion. In certain localities deficient in forests, therefore, afforestation of marginal lands, and eroded river and village wastelands should be undertaken. Forest area in excess of the indicated proportion, if any, should however, not be sacrificed. To maintain an overall average, it is essential that states better suited for the growth of trees should help to make good the deficiency in those parts where climatic and edaphic factors militate against tree growth.

3.1.16 Wildlife

The national forest policy emphasizes the need for affording protection to the animal kingdom and particularly to rare species such as the lion and the great one horned rhinoceros, which are fast disappearing. While the damage caused by such predators as wild pigs, game and porcupine cannot be denied, the elimination of their natural enemies tends to multiply them. It is necessary, therefore, that bird and animal life should be controlled by special laws and rare fauna preserved by setting up sanctuaries and large-scale national parks. For this purpose, a Central Board for Wildlife has been constituted by the government of India in the Ministry of Food and Agriculture Resolution No.7-110/51-F, dated 4th April, 1952.

3.1.17 Grazing

The controversial question of grazing in state forests calls for a clear definition of policy. Speaking generally, all grazing in forests, particularly unlimited or uncontrolled grazing is incompatible with scientific forestry. At the same time, grazing does take place in forests and must be accepted as a hard fact. There are indeed circumstances in many regions where a moderate amount of grazing does little direct harm, and may even do a great deal of indirect good in reducing the risk of fire and in suspending regression at a desirable stage. But efficient forests management requires that grazing should be regulated as regards the time and place, as also the number of cattle admitted. The formulation of the grazing policy should be based on the following cardinal principles.

(i) Continuous grazing on the same area by larger herds is destructive of the better strains of grasses and leads to a deterioration of the grass complex. Wherever it is permitted and is in great demand, efforts should be made to introduce rotational grazing, the benefits of which should be explained and demonstrated to the villagers.

(ii) Cheap forest grazing has a demoralizing effect and leads to the vicious spiral of reckless increase in the number of cattle, inadequate forest gazing, reduced quality of the herds and further increase in the numbers to offset the fall in quality. Free and indiscriminate forest grazing is, therefore, a serious disservice to cattle breeding. The notion that a farmer's wealth must be reckoned in terms of the number of cattle be owns, regardless of quality is one of the causes of India's uneconomical cattle wealth and must be combated.

(iii) Grazing should not be looked upon primarily as a source of revenue. But the simple and obvious way of regulating and controlling grazing as also improving the quality both of grazing and cattle themselves, is to institute a reasonable fee for the privilege of grazing.

(iv) Grazing must not be allowed in regeneration areas and young plantations during such periods as the seedlings require for establishment, otherwise they stand in danger of being browsed or trampled upon.

(v) Grazing incidence should be kept at a minimum in protection forests.

3.1.18 Sheep and goats

Experience gained in India and elsewhere points to the imposition of restriction on sheep grazing in forests and the total exclusion of goats therefrom. The damage to young plants caused by the browsing of these animals is often irrepairable, and their admission into the forest is incompatible with the aims and objects of forest management. The creation of special fodder reserves under strict rotational control is indicated for the purpose.

3.1.19 Shifting cultivation

The damage caused to forests by shifting cultivation in certain areas must be guarded against. To wean the aborigines, who eke out a precarious living from axe-cultivation moving from area to area, away from their age-old and wasteful practices, requires persuasion, not coercion; a missionary, not an authoritarian, approach. Possibilities of regulating shifting cultivation by combining it with forests regeneration (*Taungya*) to the benefit of both should be fully explored. Success in this direction largely depends on enlisting the cooperation of the cultivators and gaining their confidence and in showing consideration to their needs and wishes.

3.1.20 Sustained yields

With a view to conserving forest resources in perpetuity, the new forest policy requires scrupulous regard for sustained yield in the management of all classes of forest. The fluctuations in the annual out-turn of forests upset state budgets, industries, and other national enterprises, all working plans, therefore, should aim at confining them within the narrowest limits. This aspect assumes even greater significance in case where private owners manage their own forests. The compilation of all sound working plans, therefore requires

(i) the calculation of increment so that what is annually put on is annually cut, leaving the original assets intact or improved.

(ii) the preparation of the pams, and investigations on the propagation and tending of various species, their increment, the optimum conditions of their growth and the regulation of yield.

(iii) carefully planned afforestation schemes to replace inferior tree growth by valuable species of commercial importance.

Each state, therefore, should set up a permanent organization to deal with working plans their compilation and revision and deviations from them, research and statistics, as well as to conduct detailed surveys of available forest resources which are a sine qua non for a sound forest management.

3.1.21 Forest administration

The efficiency of forest administration depends directly on the adequacy of the forest laws, the training and caliber of the professional forest services, and the progress of research on both the biological and the utilization aspects of forestry.

3.1.22 Forest legislation

So far as forests under the control of the Central Government and of Part A states are concerned, adequate forest legislation exist in the Indian Forest Act and the Madras Forest Act. In some of the Part B states, there are forests regulations having the force of law. But there are some Part C states where forests laws do not exist. It is necessary, therefore, that states without a proper forest act should enact legislation at an early date on the lines of the Indian Forest Act, or validate that act for their territory. Several states have already enacted legislation for the control of private forests, it is desirable that states which have no such laws should enact them early. While framing legislation for private forests, states should not overlook the need for providing adequate staff for enforcing its provisions.

3.1.23 Forest education

Forestry courses are at present conducted for forest rangers and superior officers at the Forest Research Institute and Colleges. The states would be well advised to continue taking advantage of the facilities provided by the Central Government at Dehra Dun, associated with the well-equipped Forest Research Institute, which enjoy a world-wide reputation. A common forest education is a very effective means of inculcating an esprit de corps among officers, of developing a common outlook in forestry matters; and of ensuring concerted and integrated policies throughout the country. Openings in the profession of forestry being limited the decentralization of forest education will militate against economy and efficiency and encourage fissiparous tendencies, create unemployment, and render planned development of forest resources difficult.

3.1.24 Training of field staff

Attention has also to be directed to the proper training of lower executive staff on whose technical skill ultimately depends the proper execution of forest schemes and their extension. The tendency to start schools which are not properly equipped should be discouraged. It is necessary that contiguous states, should combine and cooperative in establishing well equipped and up-to-date training schools for the purpose of meeting their needs in the most economical manner.

3.1.25 Services

The idea held in some quarters that since forests grow by themselves, they need no technical management is based on ignorance. Inadequacy of technical personnel, and weakening of the professional standards of the men called upon to manage forests, would be followed not only by a loss of revenue but also by a general degradation of the forests, resulting in reduced output of forests produce and in deterioration of physical conditions.

3.1.26 Forest research

Investigations in the biological aspects of forestry, among others silviculture, botany and entomology, have naturally leaned heavily upon the cooperation of forest departments of various states, most of which have an organization of their own for the purpose. The maintenance of a research organization in each state commensurate with its resources and requirements in the interests of efficient forest management of the country as a whole. Research in the utilization of forest products has, on the otherhand, always been initiated and conducted at the Forest Research Institute where special equipment for the purpose exists. This balanced arrangement ensures both efficiency and economy. The institute also is naturally be the centre for specialized education in forest industries and special technical courses in paper and plywood technology, wood preservation, timber seasoning and other cognate subjects have been organized to meet the demands of industry for technicians.

3.1.27 Liaison with industry

There is also considerable scope for improvement in resecuring the utilization of the results of research on forest products by commercial and industrial interests. For ensuring closer contact between the Forest Research Institute and the interests utilizing timber and forest products, liaison and publicity arrangements at the institute need to be strengthened.

3.1.28 Popular goodwill, cooperatives and forest workers

While forest legislation, forest education, and forest research constitute the basis for sound forest management, the welfare and goodwill of the people in the neighbourhood of forests provide the firm ground on which it stands. No forest policy, however well intentioned and meticulously drawn up, has the slightest chance of success without the willing support and cooperation of the people. The recognition of their rights to forest produce at concessional rates, or, free of royalty, is not by itself enough. What is necessary is to instill in the people a direct interest in the utilization of forests. Intermediaries who exploit both the forests and local labour for their own benefit may with advantage be supplanted gradually by forest labour cooperative societies which may be formed to suit local conditions. Once the local population learns to look upon the forest as a means of its livelihood, a great step forward will have been taken.

3.1.29 Forest budgets

Forestry is a long range enterprise and it becomes incumbent upon state governments to secure for it freedom from the vagaries of the annual budget. A steady flow of funds is indispensable for sustained forest operations such as replacement of what is removed annually, improvement of remaining crops, development of communications for opening up remote areas, and protective measures; they have to be based on phased schemes which should not be set aside lightly. Budget cuts made from year to year to meet the exigencies of state finances can only be made at the expense of continuity in forest management apart from other benefits and at the sacrifice of forest revenue and other benefits in the coming year. The creation of a sinking fund, therefore, by investing a portion of the revenue in government securities, more particularly during boom years, would not only ensure availability of funds for replacement and development costs, but may also be made to act as an equalizing fund to be drawn upon in lean years to prevent a fall in revenue. The immediate profit motive should be rigidly ruled out, for, this urge may endanger the supply of large and special timber for defence and industry and lead to a disturbance of climatic conditions seriously affecting agriculture. The adoption of rotations to produce large-sized timber is often of greater importance to the general economy of the country than that of rotations which yield the maximum rate of interests in forest investments.

3.1.30 Policy an enunciation of general principles

While the discretion of state governments to regulate the details of forest administration in their respective territories is left unfettered, the general principles of the above forest policy should, in paramount national interests, be observed by them in framing their policies and legislation for the conservation of their forest resources. The forest policy of every state should be so framed as not to impinge adversely upon the general economy and physical balance of an adjoining state. It should be in consonance with the general principles underlying the forest policy laid down by the centre for the preservation and development of the nation's forest resources which are so vital to its general well-being.

3.2 National Forest Policy, 1988

3.2.1 Background and Need

The first National Forest Policy of Independent India was promulgated in 1952. However, since then many new developments had taken place and concepts of forest management changed. It was in view of the following perspectives; the need of a new forest policy was felt. The emphasis was gradually shifting from production forestry to conversion and protection of the forest resources.

The needs and attributes of the general masses with regard to forests and forestry had undergone a massive change.

1. Many new forest based industries were being set up and it was apparent that their raw material requirements could not be met from the natural forests, without upsetting the ecological balance.
2. Social, community and farm forestry had gained increasing importance in the seventies and eighties
3. There was a need to address the rights and concessions of the people and recognize the symbiotic relationship between forests and tribal
4. Provisions needed to be incorporated in the forest policy for ecological security, biodiversity conservation and compensatory afforestation.

The National Forest Policy of 1988 was issued on 7th December 1988 by the Secretary, Ministry of Environment and Forests to the Government of India.

3.2.2 Preamble

In Resolution No. 13/52/f, dated the 12th May, 1952, the Government of India in the erstwhile Ministry of Food and Agriculture enunciated a Forest Policy to be followed in the management of State Forests in the country. However, over the years, forests in the country have suffered serious depletion. This is attributable to relentless pressures arising from ever-increasing demand for fuel wood, fodder and timber; inadequacy of protection measures; diversion of forests lands to non-forest uses without ensuring compensatory afforestation and essential environmental safeguards; and the tendency to look upon forests as revenue earning resource. The need to review the situation and to evolve, for the future, new strategy of forest conservation has become imperative. Conservation includes preservation, maintenance, sustainable utilization, restoration and enhancement of the natural environment. It has thus become necessary to review and revise the National Forest Policy.

3.2.3 Objectives

The basic objectives that should govern the National Forest Policy 1988 are the following:

i. Maintenance of environmental stability through preservation and, where necessary, restoration of the ecological balance that has been adversely disturbed by serious depletion of the forests of the country.

ii. Conserving the natural heritage of the country by preserving the remaining natural forests with the vast variety of flora and fauna, which represent the remarkable biological diversity and genetic resources of the country.

iii. Checking soil erosion and denudation in the catchment areas of rivers, lakes, reservoirs in the interest of soil and water conservation, for mitigating food and droughts and for the retardation of siltation of reservoirs.

iv. Checking the extension of sand dunes in the desert areas of Rajasthan and along the coastal tracts.

v. Increasing substantially the forests cover in the country through massive afforestation and social forestry programmes, especially on all denuded, degraded and unproductive lands.

vi. Meeting the requirements of fuel wood, fodder, minor forest produce and small timber of the rural and tribal population.

vii. Increasing the productivity of forests to meet essential national needs.

viii. Encouraging efficient utilisation of forest produce and maximising substitution of wood.

ix. Creating a massive people's movement with the involvement of women, for achieving these objectives and to minimise pressure on existing forests.

The principal aim of Forest Policy must be to ensure environment stability and maintenance of ecological balance including atmospheric equilibrium which are vital for sustenance of all life forms, human, animal and plant. The derivation of direct economic benefit must be subordinated to this principal aim.

3.2.4 Essentials of Forest Management

i. Existing forests and forest lands should be fully protected and their productivity be improved. Forest and vegetal cover should be increased rapidly on hill slopes, in catchment areas of rivers, lakes and reservoirs and ocean shores and on semi-arid, arid and desert tracts.

ii. Diversion of good and productive agriculture lands to forestry should be discouraged in view of the need for increased food production.

iii. For the conservation of total biological diversity, the network of national parks, sanctuaries biosphere reserves and other protected areas should be strengthened and extended adequately.

iv. Provision of sufficient fodder, fuel and pasture, specially in areas adjoining forest, is necessary in order to prevent depletion of forests beyond the sustainable limit. Since fuel wood continues to be the predominant source of energy in rural areas, the programme of afforestation for augmenting fuelwood production to meet the requirement of the rural people.

v. Minor Forest Produce provides sustenance to tribal population and to other communities residing, in and around the forests. Such produce should be protected, improved and their production enhanced with due regard to generation of employment and income.

3.2.5 Strategies

3.2.5.1 Area under Forests

The national goal should be to have a minimum of one-third of the total land area of the country under forest or tree cover. In the hills and in mountainous regions, the aim should be to maintain two-third of the area under such cover in order to prevent erosion and land degradation and to ensure the stability of the fragile eco-system.

3.2.5.2 Afforestation, Social forestry and Farm forestry

A massive need-based and time bound programme of afforestation and tree planting, with particular emphasis on fuelwood and fodder development, on all degraded and denuded lands in the country, whether forest or non-forest land, is a national imperative.

It is necessary to encourage the planting of trees alongside of roads, railway lines, rivers and streams and canals, and on other unutilized lands under State/ corporate, institutional or private ownership. Green belts should be raised in urban/industrial areas as well as in arid tracts. Such a programme will help to check erosion and desertification as well as improve the micro climate.

Village and community lands, including those on foreshores and environs of tanks, not required for other productive uses, should be taken up for the development of tree crops and fodder resources. Technical assistance and other inputs necessary for initiating such programmes should be provided by the Government. The revenues generated through such programmes should belong to the panchayats where the lands are vested in them; in all other cases, such revenues should be shared with the local communities in order to provide an incentive to them. The vesting, in individual, particularly from the weaker sections (such as landless labour, small and marginal farmers, scheduled castes, tribals, women) of certain ownership rights over trees, could be considered, subject to appropriate regulations; beneficiaries would be entitled to usufruct and would in turn be responsible for their security and maintenance.

Land laws should be so modified wherever necessary so as to facilitate and motivate individuals and institutions to undertake tree farming and grow fodder plants, grasses and legumes on their own land. Wherever possible, degraded lands should be made available for this purpose either on lease or on the basis of a tree-patta scheme. Steps necessary to encourage them to do so must be taken. Appropriate regulations should govern the felling of trees on private holding.

3.2.5.3 *Management of State forests*

Schemes and projects which interfere with forests that clothe steep slopes, catchments of rivers, lakes and reservoirs, geologically unstable terrain and such other ecological sensitive areas should be severely restricted. Tropical rain/moist forests, particularly in the areas like Arunachal Pradesh, Kerala, Andaman and Nicobar Islands, should be totally safeguarded.

No forest should be permitted to be worked without the Government having approved the management plan, which should be in a prescribed format and in keeping with the National Forest Policy. The Central Government should issue necessary guidelines to the State Governments in this regard and monitor compliance.

In order to meet the growing needs for essential goods and services which the forests provide, it is necessary to enhance forest cover and productivity of the forests through the application of scientific and technical inputs. Production forestry programmes, while aiming at enhancing the forest cover in the country, and meeting national needs, should also be oriented to narrowing, by the turn of the century, the increasing gap between demand and supply of fuelwood. No such programme, however should entail clear-felling of adequately stocked natural forests. Nor should exotic species be introduced, through public or private sources, unless long-term scientific trials undertaken by specialists in ecology, forestry and agriculture have established that they are suitable and have no adverse impact on native vegetation and environment.

3.2.5.4 Rights and Concessions

The rights and concessions, including grazing, should always remain related to the carrying capacity of forests. The capacity itself should be optimised by increased investment, silvicultural research and development of the area. Stall-feeding of cattle should be encouraged. The requirements of the community, which determined, should be met by development of social forestry outside the reserved forests.

The holders of customary rights and concessions in forest areas should be motivated to identify themselves with the protection and development of forests from which they derive benefits. The rights and concessions from forests should primarily be for the bonafide use of the communities living within and around forest areas, specially the tribals.

The life of tribals and other poor living within and near forests revolves around forests. The rights and concessions enjoyed by them should be fully protected. Their domestic requirements of fuelwood, fodder, minor forest produce and construction timber should be made available through conveniently located depots at reasonable prices.

Similar consideration should be given to scheduled castes and other poor living near forests. However, the area, which consideration should cover, would be determined by the carrying capacity of the forests.

Wood is in short supply. The long-term solution for meeting the existing gap lies not only in increasing the productivity of forests, but also to relieve the existing pressure on forests for the demands of railway sleepers, construction industry (particularly in the public sector), furniture and paneling, mine-pit props, paper and paper board etc. substitution of wood needs to be taken recourse to. Similarly, on the front of domestic energy, fuel wood needs to be substituted as far as practicable with alternate sources like biogas, LPG and solar energy. Fuel-efficient chulas need to be popularised in rural areas.

3.2.5.5 *Diversion of forest lands for non forest purposes*

Forest land or land with tree cover should not be treated merely as a resources readily available to be utilised for various projects and programmes, but as a national asset which requires to be properly safeguarded for providing sustained benefits to the entire community. Diversion of forestland for any non forest purpose should be subject to the most careful examinations by specialists from the standpoint of social and environmental costs and benefits. Construction of dams and reservoirs, mining and industrial development and expansion of agriculture should be consistant with the needs for conservation of trees and forests. Projects which involve such diversion should at least provide in their investment budget, funds for regeneration/ compensatory afforestation.

Beneficiaries who are allowed for mining and quarrying in forest land and in land covered by trees should be required to repair and revegetate the area in accordance with established forestry practices. No mining lease should be granted to any party, private or public, without a proper mine management plan appraised from the environmental angle and enforced by adequate machinery.

3.2.5.6 Wildlife Conservation

Forest Management should take special care of the needs of wildlife conservation, and forest management plans should include prescriptions for this purpose. It is specially essential to provide for "corridors" linking the protected areas in order to maintain genetic continuity between artificially separated sub-sections of migrants wildlife.

3.2.5.7 Tribal People and Forests

Having regard to symbiotic relationship between the tribal people and forests, a primary task of all agencies responsible for forest management, including the forest development corporations should be to associate the tribal people closely in the protection, regeneration and development of forests as well as to provide gainful employment to people living in and around the forest. While safeguarding the customary rights and interests of such people forestry programmes should pay special attention to the following.

i. One of the major causes for degradation of forest is illegal cutting and removal by contractors and their labour. In order to put an end to this practice, contractors should be replaced by institutions such as tribal co-operatives, labour co-operatives, government corporations, etc. as early as possible.

ii. Protection, regeneration and optimum collection of minor forest produce along with institutional arrangements for the marketing of such produce.

iii. Development of forest villages on par with revenue villages.

iv. Family oriented schemes of improving the status of the tribal beneficiaries.

v. Undertaking integrated area development programmes to meet the needs of the tribal economy in and around the forest areas, including the provision of alternative sources of domestic energy on a subsidised basis, to reduce pressure on the existing forest areas.

3.2.5.8 Shifting Cultivation

Shifting cultivation is affecting the environment and productivity of land adversely. Alternative avenues of income, suitably harmonised with the right land use practices, should be devised to discourage shifting cultivation. Efforts should be made to contain such cultivation within the area already affected, by propagating improved agriculture practices. Area already damaged by such cultivation should be rehabilitated through social forestry and energy plantations.

3.2.5.9 Damage to Forests from Encroachments, Fires and Grazing

Encroachment on forests lands has been on the increase. This trend has to be arrested and effective action taken to prevent its continuance. There should be no regularization of existing encroachments. The incidence of forest fires in the country is high. Standing trees and fodder are destroyed on a large scale and natural generation annihilated by such fires. Special precautions should be taken during the fire season. Improved and modern management practices should be adopted to deal with forest fires. Grazing in forest areas should be regulated with the involvement of the community. Special conservation areas, young plantations and regeneration areas should be fully protected. Grazing and browsing in forests areas need to be controlled. Adequate grazing fees should be levied to discourage people in forests areas from maintaining large herds of non-essential livestock.

3.2.5.10 *Forest based Industries*

The main considerations governing the establishment of forest-based industries and supply of raw materials to them are as follows:

i. As far as possible, a forests-based industry should raise the raw material needed for meeting its own requirements, preferably by establishment of a direct relationship between the factory and the individuals who can grow the raw material by supporting the individuals with inputs including credit, constant technical advice and finally harvesting and transport services.

ii. No forest-based enterprise, except that at the village or cottage level, should be permitted in the future unless it has been first cleared after a careful scrutiny with regard to assured availability of raw material. In any case, the fuel, fodder and timber requirements of the local population should not be sacrificed for this purpose.

iii. Forest-based industries must not only provide employment to local people on priority but also involve them fully in raising trees and raw-material.

iv. Natural forests serve as a gene pool resources and help to maintain ecological balance. Such forests will not, therefore, be made available to industries for undertaking plantation and for any other activities.

v. Farmers, particularly small and marginal farmers, would be encouraged to grow, on marginal/degraded lands available with them, wood species required for industries. These may also be grown along with fuel and fodder species on community lands not required for pasture purposes, and by forest department/corporations on degraded forests, not earmarked for natural regeneration.

vi. The practice of supply of forests produce to industry at concessional prices should cease. Industry should be encouraged to use alternative raw materials. Import of wood and wood products should be liberalised.

vii. The above considerations will, however, be subject to the current policy relating to land ceiling and land-laws.

3.2.5.11 *Forest Extension*

Forests conservation programme cannot succeed without the willing support and co-operation of the people. It is essential, therefore, to inculcate in the people, a direct interest in forests, their development and conservation, and to make them conscious of the value of trees, wildlife and nature in general. This can be achieved through the involvement of educational institutions, right from the primary stage. Farmers and interested people should be provided opportunities through institutions like Krishi Vigyan Kendras, Trainers' Training/Centres to learn agrisilvicultural and silvicultural techniques to ensure optimum use of their land and water resources. Short term extension courses and lecturers should be organised in order to educate farmers. For this purpose, it is essential that suitable programmes are propagated through mass media, audio-visual aids and the extension machinery.

3.2.5.12 *Forestry Education*

Forestry should be recognised both as a scientific discipline as well as a profession. Agricultural universities and institutions dedicated to the development of forestry education should formulate curricula and courses for imparting academic education and promoting postgraduate research and professional excellence, keeping in view the manpower needs of the country. Academic and professional qualifications in forestry should be kept in view for recruitment to the Indian Forest Service and the State Forest Service. Specialized and orientation courses for developing better management skills by in service training need to be encouraged, taking into account the latest development in forestry and related disciplines.

3.2.5.13 *Forestry Research*

With the increasing recognition of the importance of forests for environmental health, energy and employment, emphasis must be laid on scientific forestry research, necessitating adequate strengthening of the research base as well as new priorities for action. Some broad priority areas of research and development needing special attention are:

i) Increasing the productivity of wood and other forest produce per unit of area per unit time by the application of modern scientific and technological methods.

ii) Revegetation of barren/margin/waste/mined lands and watershed areas.

iii) Effective conservation and management of existing forests resources (mainly natural forests eco-system).

iv) Research related to social forestry for rural/tribal development.

v) Development of substitutes to replace wood and wood products.

vi) Research related to wildlife and management of national parks and sanctuaries.

3.2.5. 14 *Personnel Management*

Government policies in personnel management for professional foresters and forest scientist should aim at enhancing their professional competence and status and attracting and retaining qualified and motivated personnel, keeping in view particularly the arduous nature of duties they have to perform, often in remote and inhospitable places.

3.2.5.15 Forest Survey and Data Base

Inadequacy of data regarding forests resources is a matter of concern which creates a false sense of complacency. Priority needs to be accorded to completing the survey of forest resources in the country on scientific lines and to updating information. For this purpose, periodical collection, collation and publication of reliable data on relevant aspects of forests to modern technology and equipment.

3.2.5.16 Legal Support and Infrastructure Development

Appropriate legislation should be undertaken, supported by adequate infrastructure, at the Centre and State levels in order to implement the Policy effectively.

3.2.5.17 Financial Support for Forestry

The objectives of this revised Policy cannot be achieved without the investment of financial and other resources on a substantial scale. Such investment is indeed fully justified considering the contribution of forests in maintaining essential ecological processes and life-support systems and in preserving genetic diversity. Forests should not be looked upon as source of revenue. Forests are renewable natural resources. They are a national asset to be protected and enhanced for the well being of the people and the Nation.

Questions

1. The first National Forest Policy of independent India is

   a) 1950 b) 1952

   c) 1955 d) 1957

2. National Forest Policy of 1988 was issued on

   a) 7th December   b) 7th December

   c) 15th December   d) 28th December

3. Forests Conservation Act was enacted in the year

   a) 1980   b) 1972

   c) 1976   d) 1950

4. The number of National Forest Policy implemented so far in our country are

   a) One   b) Two

   c) Three   d) Four

5. The apex body for forestry research and education in India is

   a) ICFRE   b) FRI

   c) IGFNA   d) IIFM

References

Anon. 1997. The Indian Forest Act, 1927 along with Forest Conservation Act, 1980 and National Forest Policy, 1988. Natraj Publishers, Dehra Dun. 175 p.

Chhtrapati Singh. 2000. Indian Forest Policy and Forest Laws. Natraj Publishers, Dehra Dun. 338 p.

Khosla, P.K. 1992. Status of Indian forestry problems and perspectives. Indian Society of Tree Scientists, Solan, H.P., India. 444 p.

Maslekar, A.R. 1981. Forester's companion. Indian Forestry Handbook, Jugal Kishore and Co. Publishing Division, Dehra Dun. 604 p.

—•—

CHAPTER 4

Introduction to Social Forestry

4.1 History

Widespread loss of tree vegetation in the developing world in the past few decades led to the emergence of social forestry. It was a response to meet the growing scarcity of biomass and to preserve the environment. Several countries in the developing world launched this type of programme to meet the demands of an expanding population for fuel, fodder and timber.

Social forestry was new approach to solve the problem of fuel scarcity with the participation of rural people who would plant, tend and maintain trees by themselves. The species planted would be of their choice. People's Republic of China was one of the first countries to embark on a major community reforestation programme. A massive nationwide campaign was launched during 1950s as a means of replenishing the country's stock of trees depleted by the previous decades of war and over exploitation. The Republic of Korea was another country where a nation wide tree growing was taken up. In 1962, Korea took the reforestation programme on a communal basis and as a major national priority.

By early seventies, the problem of deforestation and environmental degradation were expressed at International Conference at Stockholm during 1972 bringing in to sharp focus the changes required in policies towards environment and the crisis of tree depletion.

The World Congress (1978) added the dimension of forestry for the people and by the people. Food and Agricultural Organization also brought the programme of forestry for local community development.

Although it was recognized at most forums that local community forestry was the only solution to the problem of tree depletion but support, both financial and technical was negligible. It was only in the early eighties that several International organizations and agencies agreed to aid rapid tree development programme.

Social forestry was first recognized as an important component of forestry development and meeting the rural need in the Interim Report of the National Commission on Agriculture on social forestry during 1972. The commission stressed on the socioeconomic importance of social forestry for rural community as well as in the management of forest resources. It was felt that by taking up the programme of raising trees, grasses and fodder in the farmers own lands, village commons, wastelands and degraded forests close to habitations, it would be possible to meet the requirements of fuel wood, fodder, small timber for rural housing and agricultural implements etc.

In our country the concept of social forestry is not new. It is found in the preachings of Buddha about 2500 years ago. Lord Buddha preached that every good Buddhist should plant one tree and look after it over five years so that it grows to a full tree and in this way he should plant about 5 trees in his life time.

The Great Emperor Ashoka is credited to have got planted shade trees and fruit trees along the roadsides for the benefit of travelers. During early period of British rule, need for industrial expansion and communication required timber from the forests. Attempts made by the British were simply to reserve and demarcate forests for their industrial needs. No significance was attached to important role of trees to the local population.

In the monumental Report on Improvement of Indian Agriculture (1893) Voelcker observed that forests had not been preserved. His observation on keeping aside village forests for the local people was probably the first observation of importance of forests to people's economy.

Afforestation in the post independence period can be divided in to three phases. In the first phase '*Van Mahotsav*' was started in fifties which failed to attract attention largely due to ignorance at all levels. In the second phase, farm forestry was started in some states in the 1970's. The third phase was the period when social forestry programme actually took off in eighties with massive programmes and ambitious targets.

4.2 Concept of Social forestry

The word Social forestry was coined by Westoby and used in the Nineth Commonwealth Forestry Congress in 1968. According to Prasad (1985) "Forestry outside the conventional forests which primarily aims at providing continuous flow of goods and services for the benefit of people. This definition implies that the production of forest goods for the needs of the local people is social forestry. Thus, social forestry aims at growing forests of the choice of the local population.

Shah (1985) stated that conceptually social forestry deals with poor people to produce goods such as fuel, fodder etc. to meet the needs of the local community particularly underprivileged section.

4.3 Objectives of Social forestry

The objectives of Social forestry adopted by the Commission (1976) were based on the economic needs of the community aimed at improving the conditions of living.

The main objectives are

i) Fuel wood supply to the rural area and replacement of cowdung

ii) Small timber supply

iii) Fodder supply

iv) Protection of agricultural fields against wind and for recreational needs

4.4 Components of Social forestry

The components of social forestry are i) Farm forestry, ii) Extension forestry, iii) Reforestation of degraded forests and iv) Recreation forestry.

4.4.1 Farm Forestry

Farm forestry is the name given to programmes which promote commercial tree growing by farmers on their own land

Farm forestry was defined by NCA (1976) as the practice of forestry in all its aspects in and around the farms or village lands integrated with other farm operations. The Commission noted that any programme of planting in farms should serve the following objectives

i) To supplement production of fuel wood and small timber to meet the increasing requirements of local people

ii) To release cow dung for use as manure

iii) To increase production of leaf fodder

iv) To create a diverse ecosystem by having trees interspersed with cultivation

Farm forestry programme was started in seventies. Gujarat is the most widely known example. The entire component was free distribution of Eucalyptus seedlings for farmers. Initially the idea was to encourage farmers to plant on unused or marginal wastelands of their own farms. Farm forestry constituted the most important component of Social forestry in many of the states. In U.P, Gujarat, Rajasthan, Karnataka, Kerala and A.P farm forestry accounted for 70-91 per cent of the Social forestry programme.

In Tamil Nadu, two farm forestry schemes were being operated. For large farmers, seedlings were distributed at cost price. For very poor farmers, upto 500 seedlings were provided at free of cost and a small cash incentive was paid based on the number of plants surviving at the end of the first and second years. Such provision has been in practice in some other states too which contribute to raising enthusiasm among small farmers.

IFFCO, leading farm forestry cooperative initiated a project which aims at taking up 5000ha of wastelands for development in 10 states through the formation of farm forestry cooperatives.

The main objectives are to develop farm forestry on a large scale on wastelands or marginal productive lands so as to help in bringing about ecological balance, conserving soil moisture and checking soil erosion. An innovative feature of the projects is that 65 per cent cost is to be by way of bank finance and the remaining is shared equally as grant by NWDB and IFFCO.

4.4.2 Extension Forestry

Extension forestry is the practice of forestry in areas devoid of tree growth and other vegetation situated in places away from the conventional forest areas with the object of increasing the area under tree growth.

It is classified as below.

i) Mixed forestry

Mixed forestry is the practice of forestry for raising fodder grass with scattered fodder trees, fruit trees and fuel wood trees on suitable wastelands, panchayat lands and village common lands.

ii) Shelterbelts

Shelterbelt is defined as a belt of trees and or shrubs maintained for the purpose of shelter from wind, sun, snow drift, etc.

iii) Linear Strip plantations

These are the plantations of fast growing species on linear strips of land on the sides of public roads, canals and railway lines.

The National Commission on Agriculture (1976) recommended under extension forestry, the raising of plantations of quick growing species on the sides of roads, canals and railway lines. The National Social forestry Projects which were in operation in various states have taken up strip plantations. Such plantations have the capability to generate wood and other products whose utilization may help the local population to meet some of their basic requirements.

The steps recommended by NCA (1976) are that mixed forestry in the village wastelands and panchayat lands should be acceptable to people. Only quick growing and those products of immediate concern such as fodder and grass should be taken up with optimum input and technology. The programme should be undertaken only in such areas where the incidence of wastelands in a village or group of villages is sufficiently high, so that a part of it can always be kept apart for the satisfaction of the villagers rights. It was emphasized that the income from mixed forestry should be divided equally between the panchayats and state governments.

In community forestry all type of people i.e. both landholders and landless can participate. Community forestry is probably the best suited programmes for afforesting degraded wastelands especially public lands. The only effort needed is to organize people to work and cooperation with government agencies for planting and protecting trees along roadside, railway lines and canal banks.

In India, successful community forestry programmes have been launched in Gujarat, Haryana, Tamil Nadu, Kerala, U.P, Orissa, etc. Notable successes include the irrigation tank plantations in Tamil Nadu which represent the massive joint effort on the part of the state forest department and local communities.

4.4.3 Reforestation of Degraded forests

The Interim Report of NCA on Social forestry (1976) suggested reforestation of degraded forests to achieve the following objectives.

a) To grow short rotation fuel and timber species for meeting the requirements

b) To organize fuel wood supplies at reasonable rates which would prevent pilferage from neighbouring commercial forests

c) To tie up degraded forest areas with nearby rural and semi urban centres for their requirements of fuel wood

d) To provide employment

e) To rehabilitate the degraded forests

In Bihar, West Bengal and Tamil Nadu, the reforestation of degraded forest started in early fifties. Out of the total area of 75 mha, about 40 mha are reported to be degraded. The degraded area under forests needs immediate attention for ecological restoration and for meeting the socio-economic needs of the communities living in and around such areas. In Tamil Nadu, the reforestation efforts were made in the fuel wood forests of tropical dry evergreen and tropical dry deciduous forest types.

4.4.4 Recreation Forestry

Recreation forestry is the practice of forestry with the object of raising flowering trees and shrubs mainly to serve as recreation forests for the urban and rural population. This type of forestry is also known as *Aesthetic forestry* which is defined as the practice of forestry with the object of developing or maintaining a forest of high scenic value.

This aims at preservation of forests in compact blocks or development of small blocks of man made forests of miscellaneous trees offering shade and shelter, creating facilities for recreation of people such as picnic corners, zoos, facilities for hiking, riding, boating, etc. and establishment of natural history museums.

4.5 Benefits of Social forestry

The main benefits of social forestry are

i. Increase the supply of fuel wood and fodder

ii. Generate rural employment

iii. Maintain ecological balance

iv. Appropriate use of wastelands

v. Promote village and cottage industries

vi. Induce environmental and tree consciousness among people

vii. Relieve pressures from natural forests

viii. Stabilize agricultural production

Questions

1. Social forestry was first coined by

   a) Westoby b) Shah

   c) Dietrich Brandis d) None of the above

2. The practice of forestry in and around the farms/ village lands is called as

   a) Social Forestry b) Farm forestry

   c) Community forestry d) All the above

3. The practice of forestry in areas devoid of tree growth and other vegetation outside the conventional forest areas is

   a) Social forestry b) Extension forestry

   c) Afforestation d) None of the above

4. The practice of forestry with the object of raising flowering trees is called as

   a) Recreation forestry b) Urban forestry

   C) Social forestry d) Mixed forestry

5. Social forestry was first adopted successfully in

   a) Gujarat b) Tamil Nadu

   c) Kerala d) Rajasthan

6. Plantations grown as linear strips of land on the sides of public roads, canals and railway lines are called as

   a) Mixed Forestry b) Avenue plantations

   c) Linear strip plantations d) None of the above

7. Van Mahotsav' is a festival of

   a) Tree planting b) Wildlife

   c) Timbers d) None of above

8. Van Mahotsav was started in

   a) 1950 b) 1960

   c) 1965 d) 1970

References

Negi, S.S. 1986. A Hand book of Social Forestry. International Book Distributors, Dehradun.177p

Prasad, R and Bhatnagar, P. 1995. Social Forestry Experiences over a Decade. International Book Distributors, Dehradun.229p

Prasad, V.N. 1985. Principles of Social cum Community Forestry. International Book Distributors, Dehradun.108p

—•—

CHAPTER 5

Social Forestry Plantations

5.1. Railway Line plantations

As the forest resources dwindled and scarcity of fuel wood and timber in the rural areas, the Government thought of utilizing the railway lines for tree cultivation. The railway department undertook raising of plantations in railway lines (Plate 5).

5.1.1 Objectives

The objectives of planting in strips along railway lines can be stated as below:

i) Stabilization of railway track and protection of railway track against erosion

ii) Optimum utilization of the land for tree cultivation to produce products needed by the local people

iii) Checking of illicit encroachment on this land

iv) Incidental shelterbelt effect of these plantations and

v) Checking of the shifting sand in desert areas getting on to the railway track.

5.1.2 Orientation and Spacing

The safety of the railway traffic, the nature of the land available for planting, the position of the telephone and electric lines, etc. should be considered while planting along the railway lines. From the point of view of the safety, the first row of trees should not be planted very close to the railway track. The first row of trees is recommended to be planted at a distance of about 7.5 m from the centre of the track. Further, the trees in the first row should be such that it may not attain a height more than their distance from the railway track so that in the event of the wind throw the tree tops may not reach the railway line and create traffic hazard. The inner side of the curves should not be planted to ensure clear visibility of the track. On either side of an unmanned level crossing a length of about 100 m should be left unplanted. No planting should be done under the telephone line while planning orientation of tree lines.

The spacing between the lines and between the trees in a line may be 5 m, 3 m or 2 m depending upon the species to be planted, the fertility of the site and the objects of management. Commercial timber species to be harvested at comparatively longer rotations of 30 years or more may be planted at 5 x 5 m spacing. Fast growing species like eucalyptus may be planted at 2 x 2 m spacing. Leaf fodder yielding trees can be planted at close spacing of 1.5 m x 1.5 m or 2 m x 1.5 m.

5.1.3 Choice of Species

The choice of species depends on climatic and site factors and the products required to be produced from such areas. Ornamental trees should be planted near the towns, railway stations, road crossings and places of scenic beauty or of tourist interest. Trees with brittle stems or branches should not be planted especially in the first row. Wind firmness is a desirable character for the species for railway line plantation. The plantations along the railway lines are prone to accidental fires originating from engine spark especially in areas infested with tall grass or bush growth. The species to be planted in such areas should be able to resist such accidental fires.

5.2 *Canal bank plantations*

The trees along the canal bank are planted to provide comfort to the travelers using canal side roads, to stabilize canal banks and also to check encroachment on government lands acquired along the canals. A wide gap between the demand and supply of fuel wood and small timber necessitated utilization of these areas for the production of such products. The trees planted in the strips along the canals help in preventing waterlogged conditions and thereby the formation of alkali soils in adjoining lands. Trees to be planted in such areas are capable of pumping out sizeable quantity of water in view of their high transpiration rate (Plate 5).

5.2.1 Objectives

The objects of canal bank planting can be broadly stated as follows.

i) Stabilization of canal banks against erosion

ii) Checking of shifting sand getting into the canal in desert areas

iii) Utilization of the available land for tree cultivation and production of tree products, particularly fuel wood and small timber for local people

iv) Provision of comfort to travelers using canal side road

v) Checking water logging in strips along canals and in adjoining areas

vi) Shelterbelt effect on adjoining areas

vii) Improvement of aesthetics in the area.

5.2.2 Orientation and Spacing

The number of rows to be planted and the position of the first row depend upon the width and nature of the land available for planting along the side of the canal. The first row of trees is generally planted about 7.5 m from the berm of the canal in the case of multiple row planting. Subsequent rows are spaced at 5 m or 3 m or 2 m depending upon the species to be planted and the objects of management of such canal bank plantations. The spacing between the trees in a line depend upon the species selected for planting and the desired end product. The first row of trees should preferably be of shade trees. Ornamental trees may be planted in the first row or in groves near towns, villages, rest houses, picnic spots, places of tourist interest and crossings of the roads.

5.2.3 Choice of Species

For planting in the first row, the trees with a strong taproot system should be preferred. For subsequent lines the species may be selected to provide many tree products needed by the local people. The choice of species also depends on climate, site conditions, object of planting, etc. In places where the plantation strips suffer from water logging, species tolerant of water logging should be selected.

5.3 Tank Bed Plantations

Tanks are rainwater collection community irrigation reservoirs. The size of tanks range between one and several hundred hectares and are designed to supply irrigation water to a small agricultural community. These tanks are usually situated at the edge of a village and have the appearance of a large natural depression partly filled with water. Most tanks are completely empty during the driest months of the year and full for 15 -30 days at the peak of the rainy season so that the whole area of the bed is submerged. As water is let out during the dry season to irrigate the surrounding farmland, the level in the tank drops and an increasingly large area of foreshore is exposed. The tank foreshore consists of the gradually sloping sides of the tank. In many villages, these areas of temporarily exposed ground are the only substantial areas of common land on which fuel wood plantation might be established.

Afforestation of tank foreshores is the largest component of the afforestation of communal lands project. Most tank foreshores are productive sites because of the availability of moisture. Regular deposition of silt helps to maintain soil fertility.

The proximity of each tank foreshore to a village is another advantage, both labour and consumer are nearby. At the same time, the proximity to village increases the risk of damages by human being and livestock and consequently village co-operation and community involvement are essential for the creation of successful plantations and their proper protection and maintenance. Further, it is a high visibility activity with good demonstration value (Plate 5).

5.3.1 Objectives

(i) To reduce the pressure on the forest and produce fuel and fodder for village community

(ii) To convert the wastelands into economically productive lands,

(ii) To create financial resources for panchayats to take up welfare measures for community purpose

(iii) To make village panchayats also share holders in community forestry

5.3.2. Choice of species

The main species being used for planting on tank foreshores is *Acacia nilotica*. It has the essential characteristics of being capable of surviving prolonged submersion and periodic flooding. It grows quickly and produces an excellent fuel wood. It is one of the most important tree species of Social Forestry Programmes in India. It has a very wide distribution right from the semi arid regions of Rajasthan to Tamil Nadu and in eastern most part of Bihar and Orissa. It also grows in riverain beds. Every part of the tree is used in one or the other way by the villagers.

5.4 *Roadside plantations*

Avenue planting along the roads has a long history in India. The concept of avenue planting is as old as the Vedic period. The advent of the Aryans has been commonly associated with the general development of communications in the country and the planting of trees along the roads. It was, however, during the period of Ashoka and later during the reign of the Moghuls that the foundation of proper roadside avenues was laid. The Marathas also took up the construction of roads and planting of trees along road. It was, however, with the establishment of the British rule that the systematic opening of the countryside with roads was taken up. In the beginning, emphasis was laid on the construction of roads to improve the communication in the countryside and it was during the latter half of the nineteenth century that roadside avenue planting was taken up in earnest with the primary objective of providing much needed shade to the travelers, particularly during the summer months (Plate 5).

5.4.1 Objectives

The roadside plantations are required to meet the following objectives

i) Provision of comfort to the travelers
ii) Aesthetics and landscape improvement
iii) Stabilization of roadsides and checking of the shifting of sand to the road from adjoining areas in the desert,

iv) Improvement of ecological conditions

v) Maximization of the productivity of the site to meet the requirement of the local people.

The main aim of roadside planting should be the provision of comfort to the travelers. In the present stage of automobiles, planting of tree groves to provide shade and good environment at some selected places where the travelers may stop and rest is also necessary. The trees should not be as dense as it obstructs the view of the landscape on both sides of the road and to trap the exhaust gases and mist under the canopy to adversely affect visibility. Where it may be imperative to meet some demands of the local people from roadside plantations, the trees to be lopped should be planted in rows behind the first row which is planted for shade and maintain the aesthetics of the roads.

Effective roadside plantations protect the road from being covered with moving sand from the adjoining lands in arid desert areas and also from erosion along stream banks or near the bridges.

Incidentally, roadside plantations help in noise abatement. Trees and shrubs can be effectively used as a noise reducing medium. Each 30 m width of trees can absorb about 6 to 8 decibels of sound intensity. The effectiveness of a belt of trees and shrubs as noise reducing medium depends upon the height of the tree, width and overall density of planting and foliage distribution. To reduce the noise generated by high speed traffic on national highways to tolerable limits, about 20 to 30 m wide belts of trees and shrubs may be necessary. To reduce the noise generated by moderate speed car traffic in the cities, 7 to 15m wide belts of trees and shrubs may be required. Evergreen trees are better for noise abatement than deciduous trees which do not afford a barrier to sound when leafless. In hills the trees planted uphill of a highway provide maximum sound control.

5.4.2 Design of Roadside Planting

The planting systems usually followed are of the following types: viz., balanced line, unbalanced continuous line, unbalanced discontinuous line, sporadic system and parkway system. Of these, the first two are most widely practised. The balance line system produced with continuous green wall of uniform size trees. Unbalanced continuous line produced as a result of alternating avenues of different species interspersed by ornamental trees should be more desirable. It should also be ensured that thê avenues should not shut out of view the spots of scenic beauty, important land marks and places of tourist interest, lakes and canals, beautiful hill or country features and snow peaks, etc. At some selected spots the parking system of planting may be adopted to develop picnic spots or resting places for the travelers.

5.4.3 Orientation and Spacing

The position of the first row of trees and the number of rows to be planted on either side of the road depend upon the category of the road. Single row planting is normally done along village roads and other district roads while more than one row can be planted in the case of highway roads. In the case of multiple row planting the first row is generally of shade and/ or ornamental trees and the remaining rows may be planted with fast growing trees on short rotation to produce tree products to meet the requirement of the local people. The spacing of the first and also of the subsequent rows may need adjustment with respect to the position of telephone or electric lines or drains or other structures along the road. Trees should not be planted under electric or telephone lines. Bushes and short ornamental trees can, however, be planted in such places. Trees in the first row are normally planted at least 6 m away from the electric or telephone line. The spacing between the trees in the first row of shade or ornamental trees should be wider than that between the trees in the remaining rows. The crown spread of the species to be planted determines the spacing. For shade trees normally having broad crowns a spacing of 12 to 14 m is suggested. While too wide spacing may defeat the object of providing shade to the travelers, too close spacing may adversely affect the visibility. Planting of shade trees in the first row at wider spacing may result in patchy avenues as a result of the mortality. It may, therefore, be desirable to keep the initial planting closer in order to provide for mortality and selection of better trees for final retention. Planting at about half the desired spacing is, therefore advocated and when the crowns of trees begin to touch, less promising ones should be thinned to leave the remaining trees at the desired spacing. Too close spacing along hill roads particularly on the curves is dangerous. The trees should not obstruct the view at places where the pedestrians or domestic animals might usually be crossing the road near the villages.

5.4.4 Choice of Species

The choice of species is broadly determined by the climate, site conditions, object of planting, intensity of damaging agents, etc. The site conditions vary depending upon the soil and the modifications brought about in it in the course of the construction of the road. Removal of topsoil for road construction leaves only subsoil and also creates waterlogging conditions in many areas as the level of the roadside strip becomes lower than the adjoining land and road and at many places the construction of the road obstructs natural channels of drainage. Choice of the species has to be kept in view all such modifications brought about in the site. Further, the choice of species is determined by the objectives of avenue planting whether for shade, ornamental or production of tree products needed to meet the local demand.

The species should be either truly evergreen such as mango, nearly evergreen such as neem. The species with an umbrella or sub umbrella type of crown like neem, mahua, imli and mango are better than those with linear elongated crown like eucalyptus, *Polyalthia longifolia* or *Millingtonia hortensis*. The trees with heliotropic type of branching should be preferred to those with geotropic type of branching.

The species should be able to grow in the open and on soil often modified during road construction. The species should be fairly hardy, long lived, drought resistant, quick growing, wind firm and resistant to insect pests and diseases. Too rapidly growing trees like eucalypt and *Millingtonia hortensis* should, however, be avoided in the first row as these have brittle branches which may break and *disrupt* the traffic.

The species should be capable of developing a straight and clean bole up to a height of about 2 to 3 m, above which it should develop a spreading and reasonably dense crown. High crown forms ensure better visibility and are therefore ideal while those with excessively low crown spread decrease road visibility and should be avoided. The species selected should afford "collective beauty".

The species, if possible, should be of high economic value. Fruit trees should generally not be planted along the roads, but exception may be made in favour of *Mangifera indica, Tamarindus indica* and *Syzygium cumini* which form excellent avenues. Keeping in view these considerations, the species suitable for planting in the first row along the roads are the following.

a) Tropical and Subtropical Areas

Ailanthus excelsa, Albizia lebbek, Albizia procera, Anthocephalus chinensis, Azadirachta indica, Dalbergia sissoo, Ficus glomerata, Kigelia pinnata, Madhuca longifolia, Mangifera

Railway Plantation

Canal Bank Plantation

Roadside Plantation

Tank bed Plantation

Plate 5. Social Forestry Plantations

indica, Melia azedarach,Mimusops elengi, Peltophorum pterocarpum, Pithecolobium saman, Pongamia pinnata, Samanea saman, Spathodea campanulata, Syzygium cumini, Tamarindus indica, Terminalia arjuna, Terminalia catappa.

b) Subtropical and Temperate Hilly Areas

Aesulus indica, A. assamica, Toona ciliata, Cedrus deodara, Celtis australis, Cupressus sempervirens, Cupressus torulosa, Grevillea robusta, Juglans regia, Pinus roxburghii, Pinus wallichiana, Populus sp, *Salix* sp.

Questions

1. National Wasteland Development Board came into existence during the period

   a) 1980 b) 1985

   c) 1975 d) 1970

2. In India, plantation forestry was started in

   a) 1942 b) 1842

   c) 1947 d) 1847

3. The major species used for planting on tank bed is

   a) *Acacia leucophloea* b) *Acacia nilotica*

   c) *Prosopis* d) Eucalypyus

4. Which type of tree is better for noise abatement?

   a) Evergreen trees b) Deciduous trees

   c) Semi evergreen trees d) All the above

5. Heliotropic type of branching should be preferred for

   a) Roadside planting b) Canal planting

   c) Railway planting d) All the above

6. Neem and Mango are preferred for
   a) Avenue planting b) Railway line planting
   c) Canal bank plantation d) Tank bed plantation

References

Prasad, R.R. and M.P. Johagirdar. 1992. Social factors in social forestry. Chugh Publication, India. 166 p.

Tiwari, K.M. and R.V. Singh. 1990. Social forestry plantations. Oxford and IBH Publishing Co. Pvt. Ltd., New Delhi. 79 p.

Sitaram Rao, M. 1994. Introduction to social forestry. 2nd Edition. Oxford and IBH Publishing Co. Pvt. Ltd., New Delhi. 87 p.

—•—

CHAPTER 6

Social Forestry Projects and Joint Forest Management in Tamil Nadu

Social forestry project was started with the aim of taking the pressure off the reserve forests and making use of all unused and fallow lands. Under social forestry project, trees are being planted in and around agricultural fields, along railway lines and roadsides, river banks, canal banks, village common lands, government wastelands and panchayat lands. Social forestry project was implemented mainly with a financial assistance from external agencies. Some of the social forestry projects are described below.

6.1. SIDA sponsored social forestry projects

6.1.1 Social forestry project Phase I

Social forestry project phase I was launched during 1981 with the assistance from Swedish International Development Agency (SIDA) for 5 years from 1981 to 1986 and further extended upto 1987-88 with a project cost of Rs.591.38 milion.

a) Objectives

i) To increase supply of fuel wood in rural areas and to provide poles, bamboo, small timber, fodder, grass, fruit, oilseeds and other minor forest produce

ii) To establish plantations to meet the wood requirements of the villagers, and generating employment through establishing, maintaining and felling of plantations

iii) To help economically weaker sections to raise plantations in their landholdings, which are unsuitable for agriculture.

b) Components and Targets

The major components and targets in Phase I are as follows

S. No.	Components	Targets
1	Tree Cultivation Incentive Programme(TCIP)	Distribution of 12.6 million seedlings to 1 lakh families (20,000families/year)
2	Tree Cultivation Extension Programme	Distribution of 37.50 million seedlings at cost price to private farmers
3	Plantations on tank foreshores	1,32,000 ha
4	Plantations on poromboke plains	14,700ha
5	Plantations on poromboke hills	24,500ha
6	Strip plantations (roadsides, canal sides, etc.)	6,075km
7	Fodder plantations	5,250 ha

6.1.2 Social forestry Projects - Phase II

Social forestry project phase II was implemented during 1988-89 to 1992-93 with a financial outlay of Rs.854 million.

Its main objectives and components are

1. Tree Patta Scheme for landless women on poromboke and revenue lands Community Wasteland development Programme
2. Interface Forestry Programme
3. Seedling Production and Extension Programme
4. Research and Development Programme
5. Supporting Activities such as monitoring, publicity, training, co-operation with outside agencies

Components with targets and achievements in Phase II

S.No	Components	Targets	Achievements
1	TCIP	34.56 million seedlings	52.76 million seedlings
2	TCEP	90.65 million seedlings	127.83 million seedlings
3	Block plantations	145910ha	128188 ha
4	Strip plantations	7445km	8165km
5	Recreational forestry	4.8 million seedlings	0.79 million seedlings

6.2 Joint Forest Management

The recorded forest area of India is76.52 million ha which constitutes 23.28 % of the total geographical area.Out of India's population of over one billion, 360 million live in and around the forest area, exerting tremendous pressure on limited forest resources (FSI, 2000). Of these 360 million,more than 68 million are tribal/indigenous people, a large percentage of whom constitute the most disadvantaged section of society (based on per capita income, literacy rate, nutritional and health status, and lack of access to social and technical services). Of the 63.72 million hectares actual forest area, almost 40% is degraded with a crown density of less than 0.4.This degradation of forests is mainly ascribed to the rigid state control and the resulting disempowerment and displacement of indigenous tribal and hill communities accompanied by the disintegration of community based resource management.

Consequently the government of India made a major policy shift in 1988 and switched over to Joint Forest Management (JFM). JFM is an evolving policy-based program, which sets out to establish management 'partnerships' between local forest-dependent communities and the state for the sustainable management and joint benefit sharing of public forest land. To accomplish this, JFM seeks to shift the existing inequitable distribution of management control by directly involving local people and institutions in forest management. JFM does not involve the transfer of ownership over forests, but attempts instead to restructure the formal system of access, decision-making, and sharing of benefits to account for the needs of local communities. While many object to this fundamental limitation, proponents see JFM as a negotiating process, where the objectives of different users are determined through a transparent and participatory process - regardless of 'ownership' or 'property rights' alone.

Most of this area has been brought under JFM through externally aided projects as donors have made it compulsory to adopt the policy of community forestry. It is a known fact that during the last 140 years of state control, most of the land with a potential for agriculture was cleared of forest. The remaining forest in India is now concentrated in rugged and inaccessible areas and which reduces the potential to harvest and manage the forest for production purposes. The abutting population has increased manifold. Their needs have also multiplied. It is impossible to create forest-based livelihoods for all the abutting population. Consequently most of the states are following a two-pronged approach to involve communities i.e., increasing the stake of the neighbouring communities in the management and utilization of the forests, and creating alternative sources of employment to reduce the pressure on forests. But the emphasis on promotion of alternate employment is not adequate. This linking of alternate employment and forest development has been instrumental in eliciting community participation during the project phase but it does not ensure their cooperation on a permanent basis. In Tamilnadu ,a pilot JFM project was taken up in 1988 itself with financial assistance from Swedish International Development Agency (SIDA). It was tremendous success and two places namely Ariyallur and Allikulli were showcased as the success centers of JFM. But after the project was over, the forest dependents resumed wood cutting and goat grazing which brought back the area to the original degraded condition. During project phase they had got enough of wage opportunities in execution of planting and soil conservation works, which restrained them from goat grazing and illicit removal. Therefore, the only viable way of tackling the problem on permanent basis is having a multipronged approach for promotion of sustainable livelihoods so that the people's dependence on forests is permanently reduced to usufructs and indirect benefits.

6.2.1 Tamil Nadu Afforestation Project (TAP)

Tamil Nadu Afforestation Project (TAP) is funded by Japan Bank for International Cooperation with financial outlay of Rs. 500 crores. Joint Forest Management Project is being implemented in Tamil Nadu state since 1997-98. In the project, degraded forest along with abutting hamlet or habitation are selected. The forest area is divided into three zones - Lower zone or Utility zone, Middle zone or Asset creation zone and Upper zone or Eco-restoration zone. Normally, the area for all three zones is 250 ha. in which the zone-wise gap planting is taken up. The uniqueness of the project is that it has a provision of 6 lakh for village development alone under buffer zone activities over a period of three years. The aim is to reduce the dependence of villagers on forests by alternate income generation activities.

In each of the identified management units, the people's representative body called Village Forest Council (VFC) is formed which is fully involved in the planning and execution of works, protection, harvesting and benefit sharing in the management unit with focus on degraded forests. One male member and one female member from each household provided that they are willing are enrolled as its members. Any person who opts out from the membership of VFC is not entitled for any benefit. The VFC meets as and when called for but in any case at least once in three months.

Each Village Forest Council elects an executive committee in such a manner that one hamlet elects at least two members; each VFC elects a minimum of 5 and maximum of 15 members to the Executive Committee. The Panchayat (local self-government body) members of the management unit are co-opted as ex-officio members of the Executive Committee (EC). The members of Executive Committee elect its President from themselves who is also President of VFC. The Forest Ranger concerned is the member secretary of the Executive Committee who facilitates the election process of the members and President of Executive Committee. The Executive Committee is responsible for day-to-day activities of VFC and meets at least once a month. A Memorandum of Understanding is signed in the beginning between the District Forest Officer (on behalf of Forest Department) and VFC President. This gives the details of the roles and responsibilities of the Forest Department and VFC - the partners in Joint Forest Management. All the VFCs have been registered under Societies Registration Act 1975.

The Government order sanctioning the funds from JBIC soft loan gave the objective of the programme to increase the tree cover through involvement of people. But once the programme was launched it was continuously reviewed, modified, and broadened in consonance with the feedback from people and field officers. The individual grants were converted into interest free loans so that VFC can build up its corpus fund and extend loans to the remaining forest dependents for acquiring productive assets. One landmark order came from the state government for involvement of all other line departments like Public Works Department, Electricity, Health, Agriculture, Animal husbandry, Rural Development and Tribal Development, etc. for holistic development of these micro watersheds on priority basis. The involvement of other departments is monitored and guided by the state level committee headed by Chief Secretary and at the District level by the Collector. All department heads at state and district level are members of these committees. Initially slated to end in 2001-02, spurred by its success JBIC has extended the project till 2003-04.

6.2.1.1 TAP and Sustainable Livelihood Promotion

Though it was never explicitly intended even at later stages but broadening of the TAP in response to field needs has evolved it into a sustainable livelihood project from a mere JFM programme. Once the VFC is formed it is informed of the project, components and outlay. The entire decision regarding the species, community works, line department role and skill up gradation etc is taken by the VFC. This takes care of the first step of SL methodology.

a) Assets

Assets are the stocks of the resources that are available to households, which are used to pursue a combination of livelihood strategies. Improvement, enhancement and up gradation of the assets can broadly indicate the promotion of sustainable livelihoods. A macro view of the asset creation and expansion through TAP is given below.

i) *Land* : In the forest area of around 250 ha of each micro watershed 77000 saplings are planted with deliberate emphasis on non-timber forest products (NTFPs) yielding species like, *Azadirachta indica, Phyllanthus emblica, Syzygium cumini, Tamarindus indica, Bassia latifolia, Terminalia bellerica,*. In addition 25000 saplings are distributed to the villagers for growing in their own lands. The village Forest council members are given the right to collect NTFPs and a share in the final harvest. Assuming a realistic survival of 50% of saplings to become trees – there will be 38500 trees in the forest area, which will be yielding various kinds of NTFPs.This produce they can sell directly or sell after processing and value addition.

ii) *Water* : In the forest area water harvesting structures like percolation ponds and check dams are made which help in raising the water table and recharging of bore wells in the adjoining agriculture lands. The trees alone also improve the water table of micro watershed. In the state like Tamilnadu where the annual rainfall is not more than 700 mm and that too mostly in monsoon season, these water-harvesting structures have caused complete transformation of the village economy. In most of the programme areas the cropping intensity has increased and people have shifted to the cultivation of high water use crops like sugarcane and rice.

iii) *Cattle* : The forest dependents are given interest free loans from the buffer zone fund to buy high quality milk cattle.

b) Social

The formation of Village Forest Council has provided a forum to the people for coming together and exchanging views. This plateform they are using not only for JFM matters but also for taking other decisions regarding village welfare, etc.

c) Economic/Financial

Out of 24 lakhs spent in each programme micro watershed, around 15 lakhs go for wage work of planting or civil structures by the local people. This substantial wage employment available at their doorstep gives a big financial boost and enables them make full use of the infrastructure and productive assets. Apart from that the income from NTFPs goes to the VFC fund.

d) Political

Participation/empowerment: The forest villages generally constitute only a hamlet of the panchayat and therefore do not have any control or say in the panchayat affairs leaving these people absolutely meek and powerless. But the formation of VFCs has provided the real chance for their empowerment. The entire planning, execution and monitoring of the programme is done by the village Forest Councils, whose day-to-day affairs are looked after by its Executive committee headed by an elected President. The VFC President and the concerned Range Officer jointly sign all the accounts and fund applications. VFC Presidents are ex-officio members of District Level Joint Forest Management Committee, which is headed by District Collector. This imparts more prestige and glamour to the post of President. Consequently the post of VFC President has become so much coveted that people have moved up to high court to get it.

Women empowerment is given due emphasis and one third of Executive Committee members have necessarily to be women only. In each village at least one women self-help group is formed and they are encouraged to undertake economic activities individually or jointly by providing credit from buffer zone fund. The formation of VFC and women self help group have increased their collective bargaining power for assembly and parliament elections also.

e) Human

i) *Literacy :* The forest villages have high incidence of illiteracy. After adoption of this project the people have been motivated to admit their children to the schools. Wherever schools were not available, the same have been got opened through the education department or charitable organizations. In one tribal village namely Chellapanthoddi the VFC itself started a school in which 52 children who were not going to school earlier are studying enthusiastically. After three years the school has been taken over by a charitable organization.

ii) *Health*: Through the health department multispeciality health camps are conducted for a cluster of TAP villages. Siddha (Indian System of medicine) camps have also been conducted this year.

iii) *Skill up gradation*: From buffer zone funds, village youths are imparted free training to enable them become carpenters, blacksmiths, drivers, electricians, television mechanic etc., Apart from that the ladies and other elders are trained in various village level avocations like rope making, pappad making, tailoring, embroidering, beedi making etc., so that they get steady income from alternate sources and need not do wood cutting for their livelihoods.

f) Physical

There are instructions from the government to all other departments to dovetail their developmental and infrastructure activities in TAP villages on priority basis. There is a committee at the state level headed by the Chief Secretary and at the district level headed by the District Collector to guide, steer and oversee the coordination by all other departments. This has resulted in substantial improvement in the basic infrastructure like road, electricity, water, school etc., in most of the TAP villages. From the project funds also 30%buffer zone amount is spent on creation of community assets like community halls, threshing floors etc.,

6.2.1.2 Technologies

The agriculture department conducts extension camps in the TAP villages to popularize the latest package of practices for cultivation of crops, distributes seeds of hybrid varieties of various crops and popularizes the supplementary enterprises like mushroom growing, beekeeping and horticultural farming. Similarly the veterinary department conducts animal health camps and provides technical inputs for pig rearing and poultry farming. Through the small industry department guidance is provided regarding the choice of micro enterprises and various technologies available. Forest department imparts the skills for medicinal plant cultivation and agro forestry nursery raising.

6.2.1.3 Sustainability

Improvement of village infrastructure and diversification of income generation activities have enhanced the sustainability. The reclothing of the forests and water harvesting structures has reduced the vulnerability to drought, which is the most common risk. The drought proofing of these areas has gone a long way in improving agriculture income on a sustained basis. The training in alternate livelihoods has also made significant contribution in reducing the vulnerability to the natural vagaries. The forest trees are perennial assets available for generations and by their nature each successive year their yield increases and susceptibility to natural vagaries decreases. Thus restocked forests provide households a means for diversifying their subsistence and income-generating activities, optimizing their labour resources during different seasons, and "insuring" against the risks of agricultural failures.

In villages where all the components have been covered as per the above concept, the forest has been completely regenerated and there appears no reason why the people should not protect it in future also.

TAP is a massive project and it acquired SL approach in the course of implementation as the feedback from people and field officers was continuously used for broadening the project. Hence, some bottlenecks and impediments in creation of sustainable livelihoods remain. Depending upon the dedication of field officers, cooperation by district administration and interest of the villagers, all the villages covered so far, have not provided all the components of SL approach. Top down approach is not completely eliminated. As the forest area to be treated in each micro watershed is fixed irrespective of the extent of adjoining reserved forests. The number of seedlings to be planted per hectare is also constant as per the project document irrespective of the ground needs. Similarly the buffer zone amount for each watershed is 6 lakhs whereas the number of households varies. The planting on entire 250 hectares is done in first year and the buffer zone and maintenance are completed in 3 years. Consequently there is no provision for creation of alternate employment and assets after three years. Also the attention of field officers shifts to new villages and the task remains half done in the earlier villages. It will be much better if the planting is staggered for 5 years and, buffer zone and asset creation is continued for at least 10 years in each village. The cooperation of line departments need to be further enhanced and strengthened.

With the insistence of donors on community forestry approach, all the states of India have made Joint Forest Management a hallmark of their forest policy and are expanding the area under JFM through externally aided projects. But the long-term viability of JFM appears doubtful unless the issues of poverty alleviation, empowerment, sustainable development and forest regeneration are dealt together. Adoption of Sustainable Livelihood approach amply takes care of these issues. Enhancing the outlay for buffer zone activities, involving all the line departments and decentralizing more powers to local JFM institutions can broaden the existing JFM projects for promotion of sustainable livelihoods for the forest communities. Tamil Nadu Afforestation Project is a case in point. It was conceived, sanctioned and launched as a mere JFM based greening project. But the flexibility, participatory approach, continuous monitoring, review and broadening in response to field needs transformed it into a sustainable livelihood promotion project for the forest communities. It also underscores the need for present day foresters to comprehend social and economic issues in addition to the technical ones for taking up forest development successfully. However, growing of trees and promotion of sustainable livelihoods are long drawn processes hence the duration of forest development projects should not be less than 10 years.

6.2.2. National Afforestation Programme (NAP)

As per 10th Plan document of Planning Commission relating to the Forests and Environment sector, the National Afforestation Programme (NAP) Scheme was initiated by scaling-up the SGVSY project experience and converging all afforestation schemes of the 9th Plan period to avoid duplicity or redundancy, and at the same time keeping in focus the decentralization agenda of the government. NAP is being operated as a 100% Central Sector Scheme.

6.2.2.1 Objectives

The overall objective of the scheme is to develop the forest resources with people's participation, with focus on improvement in livelihoods of the forest-fringe communities, especially the poor.

NAP Scheme aims to support and accelerate the ongoing process of devolving forest protection, management and development functions to decentralized institutions of Joint Forest Management Committee (JFMC) at the village level, and Forest Development Agency (FDA) at the forest division level.

6.2.2.2. Components

Financial support under NAP Scheme is available for:

(a) Mobilisation of village JFMC, and Micro-planning in project villages

(b) Afforestation: viz., i) Aided Natural Regeneration ii) Artificial Regeneration, iii) Bamboo plantation, iv) Cane Plantation v) Mixed plantation of trees having MFP & medicinal value vi) Regeneration of perennial herbs & shrubs of medicine value and vii) Pasture Development/ Silvipasture.

(c) Soil & Moisture Conservation

(d) Entry Point Activity (for village development; average assistance Rs.1.60 lakh per village)

(e) Fencing, Monitoring & Evaluation, Training, Awareness raising,

6.2.2.3 *Institutional framework and its strengthening*

The NAP is being implemented through a 2-tier structure of Forest Development Agency (FDA) at the forest division level and Joint Forest Management Committee (JFMC) at the village level. Thus, FDA is the confederation of JFMCs in that forest division. FDAs are registered under the Societies Registration Act. JFMCs are registered either with the Forest Department or under Statutory provisions. The district-level officers of relevant line departments of the State Govt. and Panchayat Raj Institution are members of FDA.

The recent experience with present institutional framework of FDA at the forest division level and JFMC at the village level is encouraging. All States have adopted this institutional framework including in the Autonomous Hill Districts in Assam, Meghalaya and Mizoram.

The institutions of FDAs and JFMCs are highly innovative resource transfer mechanisms whereby the Govt. of India channelises funds directly to the grass root level implementing agency for the afforestation activities.

The structure of FDAs and JFMCs also caters to the gender concerns, whereby women membership to the extent of 50% has been made mandatory in these bodies. Members of Schedule Castes and Schedule Tribes are the focus group in JFMCs.

Training of FDA and JFMC members is organized by State Forest Departments, as well as by the Regional Centres of National Afforestation and Eco-development Board.

To help and guide the FDAs and JFMCs, there is also a National-level Steering Committee of NAP and another at State level called State-level Coordination committee chaired respectively by Secretary (Environment and Forests), Government of India, and Chief Secretary of the State Government.

The following additional steps are being undertaken to further strengthen these institutions:-

(a) A Training Needs Assessment exercise is being undertaken by the Regional Centres of National Afforestation and Eco-development Board to identify the needs and capacities of different categories of staff and members of JFMCs and FDAs for efficient implementation of NAP, and subsequently develop training modules for the same.

(b) It is proposed to encourage linkage of JFMCs with other rural development programme. JFMCs may leverage outside funds to augment Entry Point Activity component of NAP for this purpose.

(c) It is also proposed to formulate a financial management manual for FDAs and JFMCs so as to strengthen account keeping at the local level, and thus, empower communities for local-level decision making.

(d) A national level JFM study has also been launched to assess strengths and weaknesses of implementation of JFM, and the progress of empowerment of the local communities in decision-making.

(e) Establish a system of e-governance in National Afforestation and Eco-development Board.

6.2.2.4 *Monitoring and evaluation mechanism*

(a) With a view to increasing the efficiency of NAP to meet the above goals, the monitoring and evaluation (M&E) exercise of the FDA projects is undertaken both by the State and Central Governments.

(b) National Afforestation and Eco-development Board undertakes the first independent mid-term evaluation of the FDA projects within 24 months of sanction of the project with focus on people's participation, functioning of JFMCs and the micro-planning exercise.

(c) The final evaluation is required to be carried in the fourth year of the Project.

(d) So far, National Afforestation and Eco-development Board has launched mid-term evaluation in respect of 97 FDA projects (including 47 Projects under SGVSY Scheme of the IX Plan).

(e) In addition, the Forest Survey of India (FSI) has randomly selected 60 FDAs each in 2003-04 and 2004-05 for monitoring of area coverage and forest cover increase therein.

(f) A national level JFM study has also been launched during 2004-05 to assess strengths and weaknesses as well coverage of JFM in the country. The Draft Report has been presented before the Ministry officials on 27 Dec. 2004.

Questions

1. Social forestry project in Tamil Nadu was first launched with the assistance of

   a) SIDA b) CIDA

   c) CRIDA d) DANIDA

2. SIDA Project phase - I was implemented during

   a) 1981-82 b) 1982-83

   c) 1977-85 d) 1987-88

3. Protection and reforestation of degraded forests with the involvement of local people is

   a) JFM b) Social forestry

   c) Tamil Nadu Afforestation Project d) None of the above

4. The unit of management under JFM is

   a) Hamlet b) Village

   c) District d) Panchayat

5. The people's representation body on each of the identified management unit of JFM is

   a) VFC b) DFC

   c) NGO d) Panchayat

6. TCIP stands

   a) Tree cultivation incentive program b) Tree cultivation intensive program

   c) Tree cultivation income program d) None of the above

7. SIDA project Phase II was implemented during

   a) 1982-83 b) 1984-85

   c) 1988-89 d) 1989-90

8. Tamil Nadu afforestation project was implemented with financial assistance from

   a) Denmark b) Japan

   c) German d) Thailand

References

Prasad, R and Bhatnagar, P. 1995. Social Forestry Experiences over a Decade. International Book Distributors, Dehradun.229p

Rajiv Kumar. 1993. Issues in Social Forestry. International Book Distributors, Dehradun.173p

Sarin, M, 1995. Joint Forest Management in India: Achievements and unaddressed challenges. Unasylva 26(180): 30-36.

—•—

CHAPTER 7

Introduction to Agroforestry

7.1 History

India has been in the forefront of agroforestry research. Ever since organized research in agroforestry started worldwide about 25 years ago. Considering the country's unique land use, demographic, political, and socio-cultural characteristics as well as its strong record in agricultural and forestry research, India's experience in agroforestry research is important to agroforestry development, especially in developing nations. Agroforestry has received much attention in India from researchers, policymakers and others for its perceived ability to contribute significantly to economic growth, poverty alleviation and environmental quality, so that today agroforestry is an important part of the 'evergreen revolution' movement in the country.

Twenty five years of investments in research have clearly demonstrated the potential of agroforestry in many parts of the country, and some practices have been widely adopted. But the vast potential remains largely underexploited, and many technologies have not been widely adopted. This situation is a result of the interplay of several complex factors. The understanding of the biophysical issues related to productivity, water-resource sharing, soil fertility and plant interactions in mixed communities is incomplete and insufficient, mainly because research has mostly been observational in nature rather than process oriented. Methods to value and assess the social, cultural and economic benefits of various tangible and in tangible benefits of agroforestry are not available, and the socioeconomic processes involved in the success and failure of agroforestry have not been investigated.

On the other hand, the success stories of wasteland reclamation, and tree farming systems show that the technologies are widely adopted when their scientific principles are understood and socioeconomic benefits are convincing. An examination of the impact of agroforestry technology generation and adoption in different parts of the country highlights the major role of smallholders as agroforestry producers of the future. It is crucial that progressive legal and institutional policies are created to achieve the historical dichotomy between agriculture and forestry and encourage integrated land use systems. Government policies hold the key to agroforestry adoption.

7.2. Concept of Agroforestry

Agroforestry is a collective name for land-use systems involving trees combined with crops and/or animals on the same unit of land. It combines i) production of multiple outputs with protection of resource base; ii) places emphasis on the use of multiple indigenous trees and shrubs ;iii) particularly suitable for low-input conditions and fragile environments; iv) it involves the interplay of socio-cultural values more than in most other land-use systems; and v)it is structurally and functionally more complex than monoculture.

7.3 *Definition*

Agroforestry is any sustainable land-use system that maintains or increases total yields by combining food crops (annuals) with tree crops (perennials) and/or livestock on the same unit of land, either alternately or at the same time, using management practices that suit the social and cultural characteristics of the local people and the economic and ecological conditions of the area.

or

Agroforestry is a collective name for a land-use system and technology whereby woody perennials are deliberately used on the same land management unit as agricultural crops and/or animals in some form of spatial arrangement or temporal sequence. In an agroforestry system there are both ecological and economical interactions between the various components.

7.4 *Benefits of agroforestry*

a) Environmental benefits

i) Reduction of pressure on forest

ii) More efficient recycling of nutrients by deep rooted trees on the site

iii) Better protection of ecological systems

iv) Reduction of surface run-off, nutrient leaching and soil erosion through impeding effect of tree roots and stems on these processes

v) Improvement of microclimate, such as lowering of soil surface temperature and reduction of evaporation of soil moisture through a combination of mulching and shading

vi) Increment in soil nutrients through addition and decomposition of litter fall.

(vii) Improvement of soil structure through the constant addition of organic matter from decomposed litter.

b) Economic benefits

i) Increment in an outputs of food, fuel wood, fodder, fertiliser and timber;
ii) Reduction in incidence of total crop failure, which is common to single cropping or monoculture systems
iii) Increase in levels of farm income due to improved and sustained productivity

c) Social benefits

i) Improvement in rural living standards from sustained employment and higher income
ii) Improvement in nutrition and health due to increased quality and diversity of food outputs
iii) Stabilization and improvement of communities through elimination of the need to shift sites of farm activities.

7.5 *Constraints in agroforestry*

The following are the major constraints in agroforestry

i) Depression in crop yields due to interference effects caused by the tree
ii) Allelopathy
iii) Increased damage to crops due to birds which the trees attract
iv) Increased damage to crops due to pests for which the tree serve as alternate hosts.

7.5.1 Interference effects

In an agroforestry system, trees being the dominant partners will compete with the herbaceous substratum for resource pools of light, water and nutrients. When the immediate supply of a single necessary factor falls below the combined demands of the plant, then the competition begins. The competition is also referred as Allelospoly.

7.5.1.1 Competition for light

Many tree crops are inefficient in interception of radiant energy in their early years since full canopy formation in trees may take many years. Under such circumstances, solar radiation falling on bare soil is wasted and promotes only weed growth. Intercropping of ground cover crops with the trees in their early years will therefore help in better utilization of the resource. Reduction in crop yield from second year of tree growth under agroforestry systems has been reported.

The shade effect caused by full canopy of a 13 year old unlopped *Acacia tortilis* was considerable. Compared to open field, the total and net radiation beneath the tree canopy were only 24% and 16% respectively. Under limitation of light, there is an inevitable depression in the intercrop yield. Establishment and growth of *Cenchrus ciliaris* was poor under 13 year old unlopped *Acacia tortilis.* Forage yield of ground cover grass under *A. tortilis* plantation was also found to decline within two years of pollarding due to shading caused by newly sprouted shoots

7.5.1.2 Competition for moisture

One of the primary promises of agroforestry, especially of mixed systems rests on the assumption that trees being deep rooted abstract water from the deeper regions and therefore do not compete in the upper stratum to which the herbaceous component is restricted.The predominant occurrence of lateral roots in the top 30cm soil has also been reported in Eucalyptus and it extracted moisture mostly from the upper soil layer. Thus trees do compete with arables in the top 30cm soil profile. In intercropping experiment, depletion of soil moisture was considerable in bamboo. The other characteristics which have been associated for competition are high root density, high root-shoot ratio, high root length per unit soil volume, etc.

7.5.1.3 Competition for nutrients

A higher concentration of the fine tree roots in the soil layer up to 50cm suggests that trees also obtain most of the nutrient requirements from the soil layer up to 50cm. The lower concentration of the fine roots below 50cm soil depth suggests that the nutrient absorption from deeper soil layer may be small. The main function of the roots reaching greater depths appears to be water uptake, particularly during periods of water stress. The proportional abundance of fine roots of agricultural crops, grasses and trees suggests that there is a keen competition for nutrients between the crops and trees when grown in mixture.

Some trees are said to be great transpire of moisture. Studies carried out regarding water consumption by different species indicate that the value of water consumption per unit of biomass produced is lower in *Eucalyptus tereticornis* and *Albizia lebbeck* compared to *Acacia auriculiformis, Dalbergia sissoo* and *Pongamia pinnata* though per plant consumption of water is highest in Eucalyptus spp.

When the trees are grown together with agricultural crops either as intercrop or on bunds, their adverse effects are visible on the agricultural crops in the vicinity of the trees. The adverse effects on the growth and yield of agricultural crops are due to the competition for light, moisture and nutrients. The nature and quantum of these adverse effects depend upon i) the age and size of the trees, ii) nature of the tree species iii) nature of the agricultural crops, iv) availability of water, nutrients, light, etc. The impact of the adverse effects is greatest in the close vicinity of the trees and diminishes as the distances increases. Such effects were observed in different crops with a combination of different tree species. The effects of *Eucalyptus tereticornis, Populus deltoides, Dalbergia sissoo* and *Acacia nilotica* grown on field bunds were studied under different conditions on wheat, paddy, jowar, potato and it was found that *P. deltoides* caused the least damage to the crops but damage by the other species was higher.

7.5.2. Allelopathy

Allelopathy is the direct or indirect effect of one plant upon another through the production of chemical inhibitors that are released in to the environment.The species interaction due to chemical influences is also designated as *Allelochemistry, Phytochemical ecology or Ecological biochemistry* and *Allelobiology.*

The chemicals that impose allelopathic influence are known as allelochemicals. Allelochemicals are present virtually in all plant tissues viz., leaves, flowers, fruits, stems, roots, rhizomes, seeds, and pollen. Of these plant parts, leaves seem to be the most consistent producer of these allelochemicals. These chemicals are usually produced by plants as secondary plant metabolites.

There are four ways in which allelochemicals escape from a plant viz., i) volatilization, during which the terpenes are released from the leaves of some plant species; ii) leaching (which has shown that living or dead leaves of many plants contain growth inhibitors); iii) exudation by which roots of several crop and non-crop species release large quantities of organic compounds that inhibit the growth of other plants; and iv) decomposition, through which allelochemicals are released from plant residue.

The accumulation of tree litter on the soil under agroforestry system has deleterious effect on the agricultural crops, among which retardation of germination and growth is most common. It is reported that aqueous extracts of bark and leaf of *Acacia nilotica* had greater inhibitory effect on germination of eight arable crops viz., sorghum, cotton, cowpea, sunflower, egg plant, tomato, ladies finger and chilli and found that the inhibition of crops might be attributed due to phytotoxins present in the extracts. The allelopathic activity of commonly grown agroforestry tree species viz., *Dalbergia sissoo, Bauhinia variegata, Ficus bengalensis, Morus alba, Populus deltoides, Salix babylonica* and *Leucaena leucocephala* on germination and early growth of rice was observed.

7.5.3 Damage due to Birds

It is generally believed that planting of trees in the farm will attract the birds and thereby enhance the risk of damage for agricultural crops.

7.5.4 Trees as an alternate host to Agricultural crops

Some trees act as alternate hosts for insect pests. For instance, the jowar hopper migrates to neem tree after harvest of crop and returns to the crop after completing its life cycle on the woody perennial.

7.6 *Silvicultural options for minimizing the negative interaction*

7.6.1 Manipulation of densities and arrangement of trees

The negative effects of trees can be minimized by increasing the spacing between the trees. The studies revealed that as the distance between the hedgerows of *Luecaena luecocephala* increased, the percent reduction in crop yield decreased. The reduction in the yield of sorghum, cowpea and caster was relatively less in wider spacing compared to narrow spacing. Tree orientation is also important to reduce the negative effect of trees on crops. Planting in east -west direction may reduce the shading effect of trees on the crops.

7.6.2 Manipulation of tree crown and roots

The negative effects of trees can be further reduced by pruning of tree crown and roots. In alley cropping system, pruning of trees and applying the biomass to the soil will reduce the trees competitive ability and increase the growth and yield of the associated intercrops by providing green manure and by allowing more light to the intercrops. In situations of severe root competition for moisture and nutrients, root pruning operation or trenching may eliminate or reduce the negative effect of trees on the intercrops.

7.6.3 Choice of agricultural crops

There is a great need to identify the suitable agricultural crops which can grow well under trees with limited solar energy available.

Example:

1. Vegetables : Ginger, turmeric, potato, cucurbit
2. Agricultural crops : Cowpea, oats, maize, soybean, groundnut
3. Grasses : Cenchrus spp, Panicum,

Certainly agroforestry has considerable potential, not as the only way to improve agricultural production, but as one important way to enhance and maintain overall productivity of land is becoming more prevalent in many parts of the world.

Questions

1. The sustainable land use systems involving trees combined with crops and /or animals on the same unit of land is termed as

   a) Social forestry b) Agroforestry

   c) Mixed forestry d) All the above

2. When an immediate supply of a single necessary factor falls below the combined demand of the plant is called

   a) Allelopathy b) Allelospoly
   c) Geochemistry d) None of the above

3. The direct or indirect effect of one plant upon another through the production of chemical inhibitors is called

   a) Allelopathy b) Amensalism
   c) Commensalism d) none of the above

4. Allelochemicals produced by plants are

   a) Primary plant metabolites b) Secondary plant metabolites
   c) Leachates d) All the above

5. Allelochemicals are released by means of

   a) Volatilization b) Leaching
   c) Exudation d) All the above

6. The negative effects of trees on intercrops can be reduced by

   a) Increasing spacing between the trees b) Planting in east- west direction
   c) Pruning of branches and roots d) All the above

7. Tree competes with the neighboring crops mainly for

   a) Light b) Moisture
   c) Nutrients d) All the above

8. Allelopathy is the best example for

   a) Association b) Amensalism

   c) Parasitism d) Symbiosis

References

Chundawat, B.S. and S.K. Gautham. 1996. A text book of Agroforestry. Oxford and IBH Publishing Co. Pvt. Ltd., New Delhi. 188 p.

Dwivedi, A.P. Agroforestry - Principle and practices. Oxford and IBH Publishing Co. Pvt. Ltd., New Delhi. 365 p.

Howard A. Steppler and P.K.R. Nair. 1987. Agroforestry - a decade of developments. ICRAF, Nairobi. 335 p.

Singh, S.P. 1994. Handbook of Agroforestry. Agrotech Publishing Academy, Udhaipur. 208 p.

—•—

CHAPTER 8

Classification of Agroforestry Systems

According to Nair (1987), agroforestry systems can be classified according to the following four basis of criteria.

1. Structural Basis
2. Functional basis
3. Socio economic Basis
4. Ecological basis

8.1. Structural classification

It considers the composition of the components including spatial admixture of the woody component, vertical stratification of the component mix and temporal arrangement of the different components. The structure of a system can be defined in terms of its components and the expected role of each. In this system the type of components and their arrangement are important. Hence, Agroforestry systems can be grouped into two categories viz., a) Nature of components and b) Arrangement of components.

A) Nature of components

Based on the nature of components,, Agroforestry systems can be classified into

i) Agrisilviculture System ii) Silvipasture System iii) Agrosilvopastoral Systems and iv) Other Systems.

8.1.1 Agrisilvicultural System

Agrisilvicultural system involves the conscious and deliberate use of land for the concurrent production of agricultural crops including tree crops.

Based on the nature of the components this system can be grouped into various /practices (Plate 6).

i) Improved fallow species in shifting cultivation

ii) The Taungya system

iii) Multispecies tree gardens

iv) Alley cropping

v) Multipurpose trees and shrubs on farmlands

vi) Crop combinations with plantation crops

vii) Agroforestry fuelwood production

viii) Shelterbelts

ix) Windbreaks

x) Soil Conservation hedges

8.1.1.1 Improved Fallow Species in Shifting Cultivation

Fallows are cropland left without crops for periods ranging from one season to several years. The objective of improved fallow species in shifting cultivation is to recover depleted soil nutrients. Once the soil has recovered, crops are reintroduced for one or more seasons.

Shifting cultivation is a pattern of land use and a system of production of crops under which plots of land are cleared, cultivated for a short period for raising one, two or three crops, after which the land is allowed to rest longer than the period of cultivation. However, during the period of rest the land reverts to some modified form of its original cover. This system is practised extensively in the north-eastern hill region comprising the states of Assam, Meghalaya, Manipur, Nagaland and Tripura and the two Union territories of Arunachal Pradesh and Mizoram and to some extent in Andhra Pradesh, Bihar, Madhya Pradesh, Orissa and Karnataka states. It is called *'jhum'* in the north-eastern hill region and *'podu'* in AP and Orissa states and considered most destructive for forest areas.

The main feature of the improved fallow system of agroforestry is that trees and shrubs are not grown with crops on the same plot at the same time. The fallow periods vary from region to region but are presently becoming shorter due to an increasingly acute land shortage. The best species for the fallow system should induce good nitrogen fixation in the soil.

While the main function of the fallow is to maintain or restore soil fertility and reduce erosion, species choice should not be exclusively confined to *'soil improvers'*. Plants with marketable products should also be considered. Plants included in improved fallows should be compatible with future crops, free of any negative physical or chemical effects on the soil and not in competition with the crops to be planted later on the same site.

8.1.1.2 Taungya System

The taungya (taung = hill, ya = cultivation) is a Burmese word coined in Burma in 1850s. The taungya system was introduced into India by Dietrich Brandis in 1890 and the first taungya plantations were raised in 1896 in North

Bengal. It is practised in the states of Kerala, West Bengal and Uttar Pradesh and to a lesser extent in Tamil Nadu, Andhra Pradesh, Orissa, Karnataka and the north-eastern hill region. In southern India, the system is called *'kumri'*. It is practised in areas with an assured annual rainfall of over 1200-1500 mm.

This is a modified form of shifting cultivation in which it is permitted to raise crops in the interspaces of the forest species. The leasee or the farmer is responsible for the upkeep of a plantation. The practice consists of land preparation, tree planting, growing agricultural crops for 1-3 years, until shade becomes too dense, and then moving on to repeat the cycle in a different area. In some cases crops may be grown one year before the trees are planted. A large variety of crops and trees are grown depending on the soil and climatic conditions.

The major trees grown are *Tectona grandis,* Bamboos, *Santalum album, Tamarindus indica, Acacia nilotica, Acacia mearnsii, Ceiba pentandra,* Cashew and Rubber. The main crops cultivated are Millets, pulses, groundnut, gingelly and cotton.

Taungya systems are of three types.

(a) *Departmental Taungya:* Under this, agricultural crops and plantation are raised by the forest department by employing labourers on daily wages. The main aim of raising crops along with the plantation is to keep down weed growth.

(b) *Leased Taungya:* The plantation land is given on lease to the person who offers the highest money for raising crops for a specified number of years and ensures care of tree plantation.

(c) *Village Taungya:* This is the most successful of the three taungya systems. In this, crops are raised by the people who have settled down in a village inside the forest for this purpose. Usually each family has about 0.8 to 1.7 ha of land to raise trees and cultivate crops for 3 to 4 years.

The taungia system has the following advantages

i) Artificial regeneration of the forest is obtained cheaply

ii) Problems of unemployment are solved

iii) Helps towards maximum utilisation of the site

iv) Low cost method of forest plantation establishment

v) In every case highly remunerative to the forest departments

vi) Provision of food crops from forest land

vii) Weed, climber growth etc. is eliminated.

The taungya system has the following disadvantages.

i) Loss of soil fertility and exposure of soil

ii) Danger of epidemics

iii) Legal problems created

iv) Susceptibility of land to accelerated erosion

The taungya farmers are given the following concessions as a part and parcel of success of the system viz.,

i) Free grazing for animals;

ii) Free timber for house construction and agricultural implements

iii) Schooling facilities for children

iv) Monitory loan at nominal interest

v) Water supply through excavation of wells and construction of ponds.

8.1.1.3 Multispecies Tree Gardens

In this system of agroforestry, various kinds of tree species are grown mixed. The major function of this system is production of food, fodder and wood products for home consumption and sale for cash. Major woody species involved in this system are *Acacia catechu, Phoenix dactilifera, Artocarpus* spp., *Cocos nucifera, Mangifera indica, Syzygium aromaticum,* etc.

8.1.1.4 Alley Cropping

Alley cropping, also known as hedgerow intercropping, involves managing rows of closely planted (within row) woody plants with annual crops planted in alleys in between hedges. The woody plants are cut regularly and leaves and twigs are used as mulch on the cropped alleys in order to reduce evaporation from the soil surface, suppress weeds and/or add nutrients and organic matter to the top soil. Where nitrogen is required for crop production, nitrogen-fixing plants are the main components of the hedgerows.

The primary purpose of alley cropping is to maintain or increase crop yields by improvement of the soil and microclimate and weed control. Farmers may also obtain tree products from the hedgerows, including fuel wood, building poles, food, medicine and fodder and on sloping land, the hedgerows and prunings may help to control erosion. Alley cropping usually works best in places where people feel a need to intensify crop production but face soil fertility problems. This situation is often characteristic of crowded, densely populated areas, but may also occur wherever some farmers wish or forced to increase production on a plot of limited size.

Woody plants are introduced as hedgerows in farm fields to maximise the positive and minimise the negative effects of trees on crop management and yields. Without doubt, trees compete with agricultural crops for soil nutrients, soil moisture and light. However, the right kind of trees at the right spacing, with proper management, may actually produce a net increase in yields from farmlands. Trees may also provide new products such as fuel wood, fodder or food, in addition to the annual crops.

The position and spacing of hedgerow and crop plants in an alley-cropping system depend on plant species, climate, slope, soil condition and the space required for the movement of people and tillage equipment. Ideally, hedgerows should be positioned in an east-west direction so that plants on both sides receive full sunlight during the day. The spacing adopted is usually 4m to 8m between rows and 25cm to 2m between trees within rows. The closer spacing is generally used in humid areas and the wider spacing in sub-humid or semi-arid regions.

The position and spacing of hedgerows may also be affected by slope and the placement and design of soil and water conservation structures when these are combined with alley cropping. On sloping land, hedgerows should always be placed on the contour. If this means that they do not have the desirable east-west orientation, then they may need regular lopping or pruning of branches to prevent excessive shading of adjacent crops.

With regard to choice of species, alley cropping usually includes leguminous trees to improve soil fertility through nitrogen fixation.

An ideal alley-cropping tree species should have following characteristics.

i) It should have a sparse, small crown to permit sunlight penetration or should resprout rapidly after pruning, coppicing, pollarding or lopping.
ii) It should form a deep taproot system with few lateral root branches near the surface so as not to compete with crop roots.

iii) It should have shallow lateral roots that are easily 'pruned' by ploughing along the hedgerow, without serious damage to the plants.

iv) The leaf litter should decompose at *a* rate that makes nutrients available when they are needed in the cropping cycle.

v) Ideally, trees and shrubs used for alley cropping should fix nitrogen and should also produce wood, food, fodder, medicine or other products used by farmers or other local community.

vi) The species selected should grow well under the specific limitations of the site such as saline or acid soils, drought, flooding, heavy winds, insect pests or other hazards. The suitable species are *Cassia siamea, Leucaena leucocephala, Gliricidia sepium, Callianda calothyrsus* and *Sesbania sesban.*

8.1.1.5 Multipurpose Trees and Shrubs on Farmlands

In this system, various multipurpose tree species are scattered haphazardly or according to some systematic patterns on bunds, terraces or field boundaries.The major components of this system are multipurpose trees and other fruit trees and common agricultural crops. The primary role of this system is production of various tree products and the protective function is fencing, social values and plot demarcation. The major multipurpose trees employed in this agro forestry system are *Leucaena leucocephala, Acacia albida, Casuarina equisetifolia, Azadirachta indica, Acacia senegal, Cocos nucifera,* etc.

8.1.1.6 Crop Combinations with Plantation Crops

Perennial trees and shrubs such as coffee, tea, coconut and cocoa, are combined into intercropping systems as follows.

i) Integrated multistorey mixture of plantation crops
ii) Mixture of plantation crops in alternate or other regular arrangement
iii) Shade trees for plantation crops
iv) Intercropping with agricultural crops.

8.1.1.7 Agroforestry Fuel wood Production

In this system, various multipurpose fuel wood species are planted in or around agricultural lands. The primary productive role of this system is to produce firewood in addition to fencing, shelterbelts and boundary demarcation. Tree species commonly used as fuelwood are *Acacia nilotica, Albizia lebbeck, Cassia siamea, Casuarina equisetifolia, Prosopis juliflora, Eucalyptus tereticornis,* etc.

8.1.1.8 Shelterbelts

Shelterbelts are defined as belts/blocks consisting of several rows of trees established at right angle to the prevailing wind. The main purposes of shelterbelts are

i) to deflect air currents, ii) to reduce the velocity of prevailing winds, iii) to provide general protection to the leeward areas against the effects of wind erosion, iv) to protect the leeward areas from the desiccating effects of hot wind, v) to provide food, fodder, timber, etc.

The following are the main characteristics of shelter-belts.

a) *Shape and composition:* Shelterbelts have a typical triangular shape. This can usually be brought about by raising tall trees in the centre.
b) *Density and width:* A certain degree of penetration by wind is planned as by raising a solid wall, the protection decreases very fast on the leeward side. Shelterbelts up to 50 m width are considered ideal under Indian conditions.

c) *Height and width* : The ratio of height and width should be roughly 1:10. However, this figure may vary with local conditions.

d) *Orientation:* Orientation of shelterbelts depends on the direction and velocity of the prevailing winds. Shelterbelts may be raised in quadrangles if the wind direction tends to change very often.

e) *Length:* Length is an important consideration as far as shelterbelts are concerned. The minimum length of a shelter belt should be about 25 times its height.

f) *Choice of species: In grasses :Saccharum spontaneum, S.munja, Panicum antidotale, Cenchrus sp;* In Shrubs *:Calotropis procera, Clerodendron phlomoides, Cassia auriculata, Dodonia viscosa* ; and in Trees : *Acacia arabica, A. leucopholea, Dalbergia sissoo, Lannea coromandelica, Eucalyptus* spp., *Tamarix, articulata, Parkinsonia aculeata, Prosopis juliflora, Prosopis spicigera* and *Casuarina equisetifolia* are recommended.

8.1.1.9 Wind breaks

Wind breaks are strips of trees and/or shrubs planted to protect fields, homes, canals or other areas from wind and blowing soil or sand.

The main objectives for which wind breaks are planted include i) to protect livestock from cold winds, ii) to protect crops and pastures from hot, drying winds, iii) to reduce/ prevent soil erosion, iv) to provide habitat for wildlife, v) to reduce evaporation from farmlands, vi) to improve the microclimate for growing crops and to shelter people and livestock and vii) to retard grass fire and viii)for fencing and boundary demarcation.

When properly designed and maintained, windbreaks reduce the speed of the wind and thus its ability to carry and deposit soil and sand. They also improve growing conditions by decreasing water evaporation from soil and plants and can be used to reduce evaporation from water surfaces, such as irrigation ponds, canals or streams. In addition, windbreaks can provide wide range of useful products, viz., poles, fuel wood, fruit, fodder, fibre and mulch.

The wind breaks should be designed as follows.

a) *Permeability:* A windbreak works by filtering and breaking the force of the wind. For most purposes, permeable windbreaks which allow some wind to pass through are the most suitable. The slight movement of air through the windbreaks forms a cushion of slow moving air on both upwind and downwind sides. This deflects the main volume of wind upwards and prevents it from descending for some distance. Thus, the wind velocity in the protected area may be reduced to between 25 and 75 per cent of the wind speed.

   Dense windbreaks produce a small area of still air in a narrow strip behind the trees, but further downwind there may be considerable turbulance. However, dense windbreaks may be desirable when a high level of protection is needed for small areas such as around homesteads and work areas or for vulnerable livestock such as newborn lambs, calves, etc.

   The desired permeability can be obtained by carefully selecting tree species. Species such as *Eucalyptus* and *Casuarina* will form effective wind breaks

b) *Orientation:* For best results, windbreaks should be established at right angles to wind from which protection is needed. Windbreaks planted north-south give good results as they provide protection from wind coming from the western quarter. They also give better shading of adjacent crops and pastures than wind breaks planted east-west.

c) *Height:* The wind break height determines the size of the sheltered area. The taller the windbreak, the greater the area it protects. On level ground, a windbreak will reduce the speed of wind for about 25 times the tree height on downwind side. Maximum reduction of wind speed is in the area 5 to 15 times the tree height away from the wind break. On the upwind side some protection is gained up to a distance of 5 times the tree height away from the windbreak. Thus a wind break 20 m tall give protection from 100 m on the upwind side to 500 m on the downwind side.

d) *Length:* Wind breaks are most effective when they stretch without major gaps for distances exceeding 12 times the mature height of the trees.

e) *Number of rows:* A single row wind break should be used only where land is so valuable that only a small amount of space can be spared for tree planting. If a single row wind break is to be planted, tree species that retain their foliage to the ground and give a fairly dense growth should be selected. Eucalyptus is generally unsuitable as single row windbreaks because of their habit of losing their lower limbs. The main disadvantage of a single row is that if one tree is lost, gap is created, which reduces the efficiency of the entire wind break. Wind breaks of three to five rows are more effective for most farm situations and are less affected by gaps caused by missing trees. Tall growing species should be planted in the centre rows and small bushy species in the outside rows

f) *Tree spacing:* Distance between trees varies with the relative importance of the protective versus productive purposes of the wind breaks. Where the products of wind breaks have a high priority, then land users may favour greater number of shorter strips and a higher proportion of small trees and shrubs which provide products such as fodder and fuel wood. If the for product is for timber, the height of wind breaks and the intervals between them can be increased. When the interest is to protect valuable crops, the wind breaks should be as tall and as far apart as possible to obtain the more protection.

g) *Gaps:* Gaps are required for gates and tracks, but because of the funneling effect through gaps, wind velocity in these areas can be substantially increased. In multi row windbreaks this can be eliminated by angling the gap at about 45 degree to the prevailing wind direction. Alternatively, a few plants, trees or shrubs can be used on either side of the gate or track to broaden the gap and reduce the funneling effect. Other solutions are to plant five or six trees at an angle to the main belt as a wing or to plant a second short row to cover the gaps.

h) *Species:* In general, trees with narrow, vertical growth are ideal for wind breaks to minimize the land removed from crop production. Some fast growing species should be used to establish the desired effect as rapidly as possible. Some of the tree species used for windbreaks are *Eucalyptus, Cassia, Prosopis, Leucaena, Casuarina, Acacia, Grevillea, Syzygium, Dalbergia,* etc.

8.1.1.10 Soil Conservation Hedges

Trees can be planted on physical soil conservation works (grass strips, bunds, risers and terraces) wherein they play two roles viz., i)to stabilize the structure and ii) to make productive use of the land they occupy. In some of the sloping landscape of the country, the risers or terraces are densely planted with trees, with multiple use.

In this system the major groups of components are multipurpose trees and common agricultural species. The primary role of multipurpose trees is soil conservation and

Alley cropping

Fuelwood Plantation

Shelterbelts

Shade trees for Plantation crops

Soil conservation hedges

Windbreaks

Plate 6. Agrisilvicultural system

provision of various tree products. The following tree species viz., *Grevillea robusta, Acacia catechu, Pinus roxburghii, Acacia modesta, Prosopis juliflora, Alnus nepalensis, Leucaena leucocephala,* etc., are used for soil conservation.

8.1.2 Silvipastural System

The production of woody plants combined with pasture is known as Silvipasture system. The trees and shrubs may be used primarily to produce fodder for livestock or they may be grown for timber, fuelwood, fruit or to improve the soil (Plate 7).

This system is classified in to three categories

i) Protein bank

ii) Livefence of fodder trees and hedges

iii) Trees and shrubs on pasture

8.1.2.1 Protein bank

In this silvipastural system, various multipurpose trees especially protein rich trees are planted in or around farmlands and range lands for cut and carry fodder production to meet the fodder requirement of livestock during the fodder deficit period in winter.

The suitable trees are *Acacia nilotica, Albizia lebbeck, Azadirachta indica, Leucaena leucocephala, Gliricidia sepium, Sesbania grandiflora,* etc.

8.1.2.2 Livefence of fodder trees and hedges

In this system, various fodder trees and hedges are planted as live fence to protect the property from stray animals or other biotic influences. The main species are *Gliricidia sepium, Sesbania grandiflora, Erythrina* sp., *Acacia* sp., etc.

Plate 7. Silvipastoral system

8.1.2.3 Trees and shrubs on pastures

In this system, various tree and shrub species are scattered irregularly or arranged according to some systemic pattern to supplement forage production. The suitable species are *Acacia nilotica, Acacia leucophloea, Tamarindus indica, Azadirachta indica,* etc.

8.1.3 Agrisilvopastural System

The production of woody perennial combined with annuals and pastures is known as agrisilvopastural system.

This system is classified into two types.

i) Home gardens

ii) Woody hedgerows for browse, mulch, green manure and soil conservation

8.1.3.1 Home gardens

This system is found extensively in high rainfall areas in tropical south and south east Asia. This practice is extensively practiced in the states of Kerala and Tamil Nadu with humid tropical climates and where coconut is the main crop. Many species of trees, bushes, vegetables and other herbaceous plants are grown in dense and in random or spatial and temporal arrangements. Home gardens also support a variety of animals. Fodder grass and legumes are also grown to meet the fodder requirement of cattle.

Home gardens represent land use systems involving deliberate management of multipurpose trees and shrubs in intimate association with annual and perennial agricultural crops and livestock within the compound of individual house. The whole tree- crop - animal units are being intensively managed by family labour. Home gardens can also be called as Multitier system or home gardens are highly productive, sustainable and very practicable. Food production is the primary function of most home gardens (Plate 8).

a) Structure of Home Gardens

Home gardens are characterized by high species diversity and usually 3-4 vertical canopy strata. The layered configuration and compatible species admixture are the most conspicuous characteristics of all home gardens. Generally all home gardens consist of an herbaceous layer near the ground, a tree layer at the upper levels and an intermediate layer. The lower layer can be partitioned into two, the lowermost being at less than 1.0m in height, dominated by different vegetables and the second layer of 1.0 -3.0m height comprising fruit crops such as banana, papaya and so on. The upper tree layer can also be divided into two, consisting of emergent, full grown timber and fruit trees occupying the upper most layer of 25m height and medium size trees of 10-20m occupying the next lower layer. The intermediate layer of 5-10m height is dominated by various fruit trees.

b) Choice of species

i) *Woody species: Anacardium occidentale, Artocarpus heterophyllus, Citrus* sp, *Psiduim guajava, Mangifera indica, Azadirachta indica, Cocus nucifera,*

ii) *Herbaceous species:* Bhendi, onion, cabbage, pumpkin, sweet potato, banana, beans, etc.

8.1.3.2 *Woody Hedgerows*

In this system various woody hedges, especially fast growing and coppicing fodder shrubs and trees are planted for the purpose of browse, mulch, green manure, soil conservation, etc. The following species viz., *Erythrina* sp, *Leucaena luecocephala, Sesbania grandiflora* are used.

Plate 8. Home Gardens

8.1.4 Other Systems

8.1.4.1 Apiculture with trees

In this system, various honey (nectar) producing trees frequently visited by honeybees are planted on the boundary of the agricultural field. eg. *Azadivachta indica, Syzygium umini, Glyricidia sepium,* etc.

8.1.4.2 Aqua forestry

In this system, various trees and shrubs preferred by fish are planted on the boundary and around fish ponds. Tree leaves are used as forage for fish. The main role of this system is fish production and bund stabilization around fish ponds. e.g. *Leucaena leucocephala, Morus alba,* etc.

8.1.4.3 Mixed wood lots

In this system special location specific multipurpose trees (MPTs) are grown mixed for various purposes such as wood, fodder, soil conservation, soil reclamation, etc.

B) Arrangement of components

Based on the arrangement of components, Agroforestry systems are classified into two categories viz., Spatial arrangement and ii) Temporal arrangement.

Spatial arrangement of plants in an agroforestry mixture may vary from dense mixed stands (e.g. Homegardens) to sparsely mixed stands (e.g. Silvopastoral system). Moreover, the species can be in zones varying widths. There can be several forms of such zones varying from microzonal arrangements (such as alternate rows) to macro zonal ones. A typical example of the zonal pattern is hedgerow intercropping (alley cropping). An extreme form of zonal planting is the boundary planting of trees on edges of plots and fields for a variety of purposes and outputs (fruits, fodder, fuel wood, fencing and protection, soil conservation, windbreak, etc,).

Temporal arrangement of plants in agroforestry can alsc take various forms. An extreme example is the conventional shifting cultivation cycles involving 2 to 4 years of cropping followed by more than 15 years of fallow cycle when a selected woody species or mixture of species is planted or is allowed to regenerate naturally. Similarly, some silvopastoral systems may involve grass leys in rotation with woody species, with the some species of grass remaining on the land for several years during the grass phase. These temporal arrangements of components in agroforestry have been described by terms such as coincident, concomitant, overlapping, separate and interpolated.

8.2 Functional Classification

Production and protection are two fundamental attributes of all agroforestry systems This implies that agroforestry systems have two functions viz.,i) Productive function yielding one or more products include food, fodder, fuel wood, NTFP, etc., and ii) Protective functions include windbreak, shelterbelt, soil conservation, moisture conservation soil improvement , shade, etc.,

8.3 Socioeconomic classification

Socioeconomic criteria such as scale of production and level of technology input and management, have also been used as a basis for classifying agroforestry systems. This is grouped into commercial, intermediate and subsistence systems.

The term commercial used when the major aim of the system is production of the output for sale. In these systems, the scale of operation is often medium to large and land ownership may be government, or corporate or private; labour is normally paid or otherwise contracted. Examples, include commercial production of agricultural plantation crops such as rubber, oil palm and coconut, with permanent under storey of food crops, or integration of pasture and animals; commercial production of shade tolerant plantation crops like coffee, tea, and cocoa under over storey shade trees; rotational timber/food crops systems in which a short phase of food crop production is used as a silvicultural method to ensure establishment of the timber species; and commercial grazing and ranching under large scale timer and pulp plantations.

"Intermediate" agroforestry systems are those that are intermediate between commercial and subsistence scales of production and management, i.e., production of perennial cash crops and subsistence crops undertaken on medium to small sized farms where the cash crops satisfy cash needs, and the food crops meet the family's food needs. Usually farmers who either own the land, or have long term tenancy rights to land, reside and work on the land themselves, and are supplemented by paid temporary labour. The main features distinguishing the intermediate system from the commercial system at one end and from the subsistence system on the other are holding size and level of economic prosperity. Several agroforestry systems in many parts of the world can be grouped as intermediate systems, especially those based on a large number of fruit trees, especially in the Asia-Pacific region and short rotation timber species such as *Albizia falcataria* in the Philippines and Indonesia

Anthropologists define subsistence farmers as those who produce most of what they consume, or consume most of what they produce. Farmers who do not, or cannot, produce enough for the needs of their families. Subsistence agroforestry systems are those where the use of land is directed towards satisfying basic needs and is managed by the owner or occupant and his/her family. Cash crops, including the sale of surplus commodities, may well be part of these systems, but are only supplementary. Most of the agroforestry systems practiced in various parts of the developing countries come under the subsistence category. Forms of traditional shifting cultivation found throughout the tropics are the most widespread example.

8.4 Ecological classification

Most of the agroforestry system characterizations pertain to specific ecological conditions of different geographical regions. It is thus easy to find several descriptions of agroforestry systems i.e. the highland, sub humid tropics or the tropical highlands. Similarly, a large number of system descriptions can be found for other ecological regions. Descriptions of existing system, as well as recommendations of potential agroforestry technologies, for specific agroecological zones, include a mixture of various forms of agroforestry; there can be agrisilvicultural, silvopastoral or agrosilvopstoral systems in any of the ecological regions.

The main point is that several types of agroforestry systems and practices are relevant to any major agroecological zone; depending on the special conditions of a zone, the emphasis of the system or practice will also vary. For example, in the tropical highlands, one of the main considerations would be the protective role of agroforestry, whereas in sparsely-populated, semiarid savannas, silvopastoral systems would be more common.

Questions

1. The broad and well known classification of agroforestry systems was given by

   a) Nair b) Dwivedi

   c) Chundawat d) None of the above

2. A common example of the zonal pattern is

   a) Hedgerow intercropping b) Home gardens

   c) Silvipastural system d) None of the above

3. The fundamental attributes of all agroforestry systems are
   a) Production and protection b) Production alone
   c) Protection alone d) None of the above
4. Integration of trees with grasses is known as
   a) Silviculture b) Agrisilviculture
   c) Silvipasture d) None of the above
5. Taungya was first introduced by
   a) Dr. Brandis b) Chaturvedi
   c) L. S. Khanna d) None of the above
6. When the use of land is directed towards satisfying basic needs and is managed by the owner is said to be
   a) Subsistence agroforestry system b) Commercial agroforestry system;
   c) Intermediate agroforestry system d) All the above
7. Jhum cultivation is also called
   a) Shifting cultivation b) Taungya
   c) Both of the above D) None of the above
8. The deliberate use of land for the production of crops including trees and crops is called as
   a) Agrisilviculture b) Silvihorticulture
   c) Agrisilvipasture d) None of the above
9. The main aim of raising crops in departmental taungya is to
   a) Reduce the weed growth b) To provide additional income
   c) Maximum utilization of site d) All the above

10. The practice of managing rows of closely planted woody plants with annual crops planted in alleys in between hedges is called as

    a) Hedge row inter cropping b) Taungya system

    c) Agroforestry fuelwood production d) Crop combinations with plantation crops

11. Shelter belts should possess a typical

    a) Rectangular shape b) Square shape

    c) Triangular d) Linear shape

12. Strips of trees planted to protect fields, homes, canal from wind is referred to

    a) Shelter belts b) Wind breaks

    c) Green belts d) None of the above

13. Planting nectar producing trees on the boundary of the agricultural field is called as

    a) Apiculture with trees b) Aquaforestry

    c) Boundary planting d) None of the above

References

Nair.P.K.R. 1993. Introduction to agroforestry. Kluwer Academic Publishers, Netherlands. 499p.

Chundawat, B.S. and S.K. Gautham. 1996. A text book of Agroforestry. Oxford and IBH Publishing Co. Pvt. Ltd., New Delhi. 188 p.

Dwivedi, A.P. 1988. Agroforestry - Principle and practices. Oxford and IBH Publishing Co. Pvt. Ltd., New Delhi. 365 p.

Howard A. Steppler and P.K.R. Nair. 1987. Agroforestry - a decade of developments. ICRAF, Nairobi. 335 p.

—•—

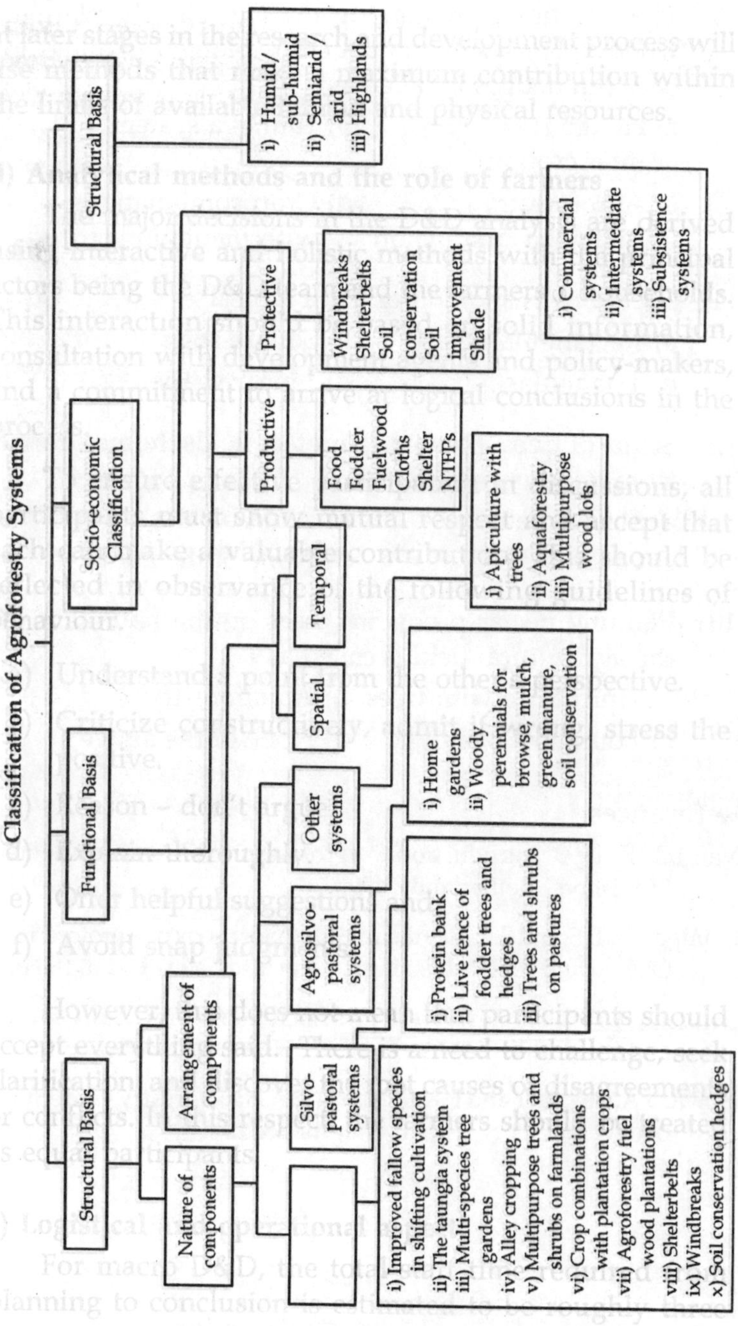
Classification of Agroforestry Systems
Structural Basis
Functional Basis
Socio-economic Classification
Structural Basis
i) Humid/ sub-humid
ii) Semiarid / arid
iii) Highlands
Nature of components
Arrangement of components
Productive
Protective
Food
Fodder
Fuelwood
Cloths
Shelter
NTFPs
Windbreaks
Shelterbelts
Soil conservation
Soil improvement
Shade
i) Commercial systems
ii) Intermediate systems
iii) Subsistence systems
Spatial
Temporal
Silvo-pastoral systems
Agrosilvo-pastoral systems
Other systems
i) Improved fallow species in shifting cultivation
ii) The taungia system
iii) Multi-species tree gardens
iv) Alley cropping
v) Multipurpose trees and shrubs on farmlands
vi) Crop combinations with plantation crops
vii) Agroforestry fuel wood plantations
viii) Shelterbelts
ix) Windbreaks
x) Soil conservation hedges
i) Protein bank
ii) Live fence of fodder trees and hedges
iii) Trees and shrubs on pastures
i) Home gardens
ii) Woody perennials for browse, mulch, green manure, soil conservation
i) Apiculture with trees
ii) Aquaforestry
iii) Multipurpose wood lots

CHAPTER 9

Integrated Farming System

Integration of various farm enterprises in farming system such as crops, vegetables, fruits, flowers cultivation, dairy, poultry, fish, goat, pig, sericulture, mushroom cultivation, agroforestry, bee keeping, silviculture, agro-based industries and food processing in a farm ensures growth and stability in overall productivity and profitability. Also, it ensures the recycling of residues, optimization of the resources and minimization of risk and generation of employment. A judicious mix of enterprises complementary to cropping and suited to the given farm situation and farmers preference would bring overall prosperity.

9.1 Why we need IFS?

With India's population crossing one billion, advent to virtual stagnation in cultivable area, slow pace in expansion of irrigated area, declining trends of food grains reserve position in the world and the plateauing tendency of the crop-response curve to progressively intensive use of agricultural inputs,the country has, resorted to and continue to pursue the pragmatic path of ensuring food security for sustainable agricultural paradigm. To meet the demand of ever increasing population growing at the rate of 1.8 per cent per annum, we would need 260-264 million tonnes of food grains by 2030. Besides, 130-152 million tonnes of milk, 151-193 million tonnes of vegetables, 84-106 million tonnes of fruits, 10-14 million tonnes of meat, 550 million eggs, 10-14 million tonnes of fish and 12 million tonnes of edible oil would be required to provide adequate nutrition to a population of 1.3 billion people. The present situation would be further compounded in view of the shrinking arable land and depleting natural base. To meet the needs of growing population, agricultural sector should have the objective of achieving increase in production resource efficiency, conservation of natural resources and human resource development. To increase productivity and growth for ensuring better standards of living and to eradicate poverty hunger and malnutrition and the sustainability of food production could be achieved through increased total productivity per unit area in a specified time.

The traditional cropping pattern in India is subject to a high degree of risk and uncertainty in yield and provides only seasonal, irregular and uncertain income and employment to the farmer. Hence, the reality of integration could only be achieved by IFS through residue recycling from one component over the other and attaining maximum net return in a holistic approach. The IFS approach therefore assumes greater opportunity for better management of farm resources to enhance the farm productivity, reduce environmental degradation and improve the quality of life of resource poor dry land farmers and to maintain the economic sustainability. Hence, the concept of vertical expansion of plant components to capture maximum solar energy for increasing the biomass production through various alternate land use systems becomes imperative. This systems help in efficient utilization of resources like land and labour. To meet out the requirement of 400 million bovines in India, there is a scarcity of green fodder than the dry fodder under dry land situations. Inclusion of perennial drought tolerant fodder grasses along with goat rearing not only rescue the farmers but also maximize the income of the resource poor dry land farmers. In India 76.2 per cent of farming community belongs to small and marginal farmers. More than 80 million farm holdings in India are below the size of 1.0 ha leaving no scope for expanding the individual farm size because of steady explosion of population with shrinkage of cultivated land due to urbanization and industrialization. The income from seasonal field crops on small and marginal farms is hardly sufficient to sustain the farmer's family. With the gradual decline in farm size, it would be increasingly difficult to produce enough food for the farm family in coming years. Under such circumstances it will be required to undertake some land based enterprises, which will complement their existing farming activity to get more employment, leading to better standard of living. Such enterprises include livestock, fishery, poultry, duckery, apiary, mushroom production and tree farming, etc.

9.2 Concept of IFS

In IFS, all the activities interact with market forces and ecosystem for purchasing inputs and disposing outputs by utilizing and degrading natural resources (land, water, air, sunshine, etc.). Sustainability is the objective of the farming system where production process is optimized through efficient utilization of inputs without infringing on the quality of environment with which it interacts. In the integrated farming, the yield would be inherently more sustainable, because the wastes of one enterprise become the input of another leaving almost no waste to pollute the environment.

9.3. Components description

9.3.1 Quail farming

Japanese quail farming is very popular as these birds do not require any vaccination. Quail farming requires less space, the reproduction is fast and the birds are resistant to poultry diseases and they fetch better returns. The Japanese quail is a tiny domesticated inter-breeding avian species, a prolific producer of meat and eggs. Adults weight 250 to 300 g and are easy to maintain. The life cycle is very short, with the female quails starting the egg laying (sexually maturing) at 7 to 8 weeks of age. Incubation is 17 days. Every adult male conveniently mates with 3 to 4 adult females, both in cages as well as on floor, to produce fertile eggs with an average hatching capacity of 70 to 75 per cent. Quails produce 300 eggs or more eggs per year. The rapid growth rate makes quails an ideal meat bird. From 6 to 10 g at day-old, the quail can reach a body weight of 175-220 g in 5-6 weeks. The quail's double role as egg and meat yielder makes quail farming a viable commercial poultry enterprise. These factors make quail farming an attractive proposition all over the country.

These birds need slightly higher levels of protein, energy and vitamins than chicken. A level of 27 per cent protein and 2800 k cal ME/kg during 4-5 weeks of growing period. The adult quails should be fed ration containing 22 per cent protein and 2900 k cal ME / kg. The quail eggs had albumen and yolk ration of 61:39.The percentage of yolk is more in quail eggs. Considering the production potential and greater acceptability, quail production has increased significantly all over the country.

Quail meat is tender, tasty, nutritious and increasingly popular as a delicacy. Meat from five week old quails has 74 per cent moisture, 20.5 per cent protein, 2.8 per cent ether extract and 1.2 per cent total ash. With these characters the quail can be effectively integrated in the conventional cropping system to enhance the productivity and net income from the farm.

9.3.2 Integrated duck farming

Ducks are the second largest source of table eggs in the country. There are about 23 million ducks in India. West Bengal has the highest duck population followed by Assam, Tamil Nadu, Andhra Pradesh, Bihar, Kerala, Orissa, Jammu and Kashmir and Tripura. Ducks predominantly are of indigenous type kept for egg production on natural foraging and have a production potential of about 130-140 eggs/bird/year. Duck eggs even have a preference over chicken eggs in certain states. Considering the importance and scope of duck rearing in our rural economy, technologies have been evolved for giving a big push to duck farming. Improved strains of ducks are now available in the country.

9.3.3 Integrated turkey farming

Turkey meat is among the most relished white meat apart from broilers and pork in the western countries. It is highly integrated in North America, Great Britain and many other European countries. Turkey farming is practiced experimentally in Haryana Agricultural University, Hisar University of Agricultural Sciences, Bangalore and Rajasthan College of Agriculture, Udaipur.

Turkeys originated in Mexico and North America. It is however, interesting to note that present turkey industry in America developed from the domesticated turkeys imported from Europe and not from the native stocks. A black beard attached to the upper breast region in case of male and it is small in female. A fleshy protuberance called snood is present near the base of the beak in both males and females. Typical male (tom) and female (hen) breeders of large turkey at 22-26 weeks of age attain a weight of 10-11 kg and 6-8 kg respectively. Adult weight of toms and hens ranges from 8 to 9 kg and 4 to 7 kg respectively in small type turkeys. Toms are marketed at the age of 17.5 weeks on reaching a weight of 12 kg and market age for hens is 14.5 weeks at an average weight of 12 kg. In spite of their huge size, turkeys have a very nervous temperament. They have very poor sense of protecting themselves.

Turkeys have higher growth rate compared to chickens, hence they require relatively higher feed and nutrient requirements. Adults require 14 per cent crude protein and 2900 k cal / kg metabolizable energy in the feed. Large turkeys raised to 14 kg weight till 24 weeks of market age consume 1.7 kg feed / 0.4 kg of live weight gain. Small turkeys are more efficient converters of feed to live body weight. Turkeys can be fed all dry feed comprising pellets, wet mash, grain and green feed in well balanced combination.

9.3.4 Integrated pigeon farming

Domestic pigeon can be reared for its meat. Females are smaller than the males and finer in the head. They hold their trails higher than males and their pelvic pones are spaced wide part. The males are large, aggressive and make a loud cooling noise. They strut around females during the breeding season.

9.3.5 Integrated poultry farming

Poultry are one of the fastest growing food industries in the world. Poultry meat accounts for about 27 per cent of the total meat consumed world wide, and its consumption is growing at an average of five per cent annually. Poultry industry in India is relatively a new agricultural industry. Till 1950, it was considered a back yard profession in India. In the sixties, the growth rate of egg production was about 10 per cent and it increased to 25 per cent in the seventies. The growth rate came down to 7-8 per cent by 1990 due to price rise in poultry feed. Broiler production is increasing at the rate of 15 per cent per year. It was 31 million in 1981 and increased up to 300 millions in 1995. The average global consumption is 120 eggs per person per year and in India, it is only 32-33 eggs per capita per year.

Specific poultry stocks for egg and broiler production are available. A majority of the stocks used for egg production and crosses involving the strains or inbred lines of white leghorn. To a limited extent, other breeds like Rhode Island Red, California grey and Australopo are used. Heavy breeds such as white Plymouth rock, white Cornish and New Hampshire are used for cross breed at the time of purchase.

Adequate space should be provided for the birds. Floor area of about 0.2 m^2 per adult bird is adequate for light breeds such as white leghorn. About 0.3-0.4 m^2 per bird is required for heavy breeds. The house should have good ventilation and reasonably cool in summer and warm during winter. It should be located on well drained ground safe from flood water.

The feed conversion efficiency of the bird is far superior to other animals. About 60-70 per cent of the total expenditure on poultry farming is spent on the poultry feed. Hence, use of cheap and efficient ration will give maximum profit. Ration should be balanced containing carbohydrates, fats, minerals and vitamins. From the day old to four weeks of age, birds are fed on starter ration and thereafter finisher ration, which contain more energy and 18-20 per cent protein. Feed may be given 2-3 times. In addition to the food stuffs, antibiotics and drugs may also be added to the poultry ration. Riboflavin is particularly needed.

The chick must be vaccinated against Ranikhet diseases with F1 strain vaccine within the first 6-7 days of age. When chicks get the optimum body weight of 1.0-1.5 kg around six weeks, they can be marketed for broiler. Hens may be retained for one year for production i.e. upto the age of about 1½ years. After that they are disposed of for table purpose. One hen is capable of laying 180-230 eggs in a year starting from the sixth month. In addition, a lying hen produces about 230 g of fresh droppings (75% moisture) daily.

9.3.6 Integrated rabbit rearing

Rabbits are reared all over the world in various climatic conditions. Rabbit rearing in India is of recent origin, though hunting of wild rabbits for meat is not uncommon. Rabbit can easily be reared with relatively with very high reproduction rates. Rabbit meat is pure white and is low in fat and cholesterol content. Further rabbit meat does not have any religious taboo, unlike pork and meat from cow.

Rabbits can be classified into two categories as meat and wool producing ones. Important meat breeds are New Zealand white, Soviet Chinchilla and Kandinrex. Breeds for wool are Angora, polish Bolamiona, Dan and Florida.

Rabbits can be kept in huts outside or in cages inside. Cages should have door (30 x 30 cm) with welded wire mesh at floor on the sides. The area required by a rabbit depends on the body weight. For one kg weight, 1.8 sq.m area is needed. Angora rabbits cannot be kept at high temperature and should be restricted to hills. Rabbit keeping requires calm and quite areas. Sudden noise can cause death of rabbits. Disturbance by dogs, cats and human beings should be kept to the minimum.

Rabbit is a non-ruminant herbivorous animal having a simple stomach. Like any other animal, rabbit requires protein, carbohydrates, fats, minerals and nutrients in its feed. Rabbits digest the fibre fraction very poorly and a diet having 5-7 per cent crude fibre is adequate. Alfalfa and other forages are well utilized by rabbits. The feed ingredients may constitute maize (20%), alfalfa (25%), wheat bran (40%), groundnut cake (12%), mineral mixture (2%) and salt (1%). The feed mixture should be given twice a day, in the morning and the evening. Other grasses and vegetables can be given intermittently in small quantities. Water should be provided.

Rabbits of 5-6 months age are ready for breeding and the body weight should not be less than 2.0 kg. The female rabbit is taken to the cage of male for mating. At the time of mating, the male produces a particular noise. After that the female rabbit is taken out of the cage. Early morning is the best time for mating. Rabbit mostly breed from March to July. Gestation period is 28-38 days. The eyes, if the young ones (pups) open around two weeks and at three weeks, they feed themselves. During this period, the pup should not be often handled. Around four weeks, the young ones are separated from the mother. After two weeks of the weaning, the female can be allowed for mating.

In a year, it gives birth five times, each time producing about six numbers. Mortality in the kindlings is the most important component that determines the economics of rabbit rearing. In general, mortality in adult rabbit is very low ranging from 3-5 per cent. However, it ranges 10-15% in kindlings up to the age of 30-42 days. Mating during extreme hot and cold season causes high mortality of kindlings.

9.3.7 Integrated piggery farming

Pigs are maintained for the production of pork. They are feed with non-edible feeds, forages, certain grain byproducts obtained from mills, meat byproducts, damaged feeds and garbage. Most of these feeds are either not edible or not very palatable to human beings. The pigs grow fast and are prolific feeder, farrowing 10-12 pigs at a time. It is capable of producing two litters per year under good management conditions. The carcass return is high at 65-79 per cent of the live weight.

Improved breeds of large white pork shire and landrace are being used widely. Pork shire is the most extensively used exotic breed in India. It is a prolific breed having good carcass quality, growth rate and feed conversion ability. A good boar weighs 90 kg in about 5-6 months and is strong on feet and legs. The mother of the pig to be selected should have large litters of eight piglets or more.

Housing should be provided with maximum comfort to pigs so that their growth is optimum. There should not be dampness and overheating. Locally available materials can be used for housing. Eight boars can be kept in 2.7 – 4.5 sq.m. areas.

Feeds play an important role in successful pig production. Pigs are the most rapidly growing livestock and suffer more from nutritional deficiencies than the ruminants. Proteins, carbohydrates, fat, minerals, vitamins and ample good water form a complete diet for pigs. Pigs have a simple stomach. Therefore they must feed with maximum of concentrates and minimum of roughage. The main ingredients of swine ration are cereals and millets and their byproducts. The fibre content in the swine ration should be very low (around 5-6%) for better feed utilization efficiency. Mixed ration should also contain 0.5 per cent of added salty. Swine requires comparatively high percentage of Ca and P than do cattle or sheep. When pigs are maintained with agricultural / kitchen waste or fish and slaughter house waste, the cost of production remains low. On an average, the consumption of feed is 3.5 per cent of total body weight.

As a general rule, well developed gilts weighing about 100 kg, when 12-14 months old, may be used for breeding. The body weight is more important than age at breeding. Sows with low body weight show higher rate of fetal and pre-weaning mortality and have been proved as mothers with poor nursing ability. The gestation period is on an average 114 days. Litter size at birth may be 1-16 numbers with the body weight of 1-25 kg. Normal period between birth of piglets is 10-20 minutes. Time taken for the whole process ranges from 1.5 to 4.0 hours. Sows are weaned after 40 days. It is advantageous to cull the sow after fifth or sixth litter in a commercial herd. Weaned sows come into heat in 3-10 days after weaning and may be allowed to breed. The boar-sow ratio should be 1:15. It is profitable to raise two litters from each sow each year.

Inclusion of pigs as subsidiary activity is gaining importance currently. It has good potential for more income provided care is taken for housing and feeding the animals. Availability of adequate water and good marketing facilities are essential.

9.3.8 Integrated goat farming

In India, activity of goat rearing is sustained in different kinds of environments, including dry, hot and cold, high mountains or low lying plains. Goat is mainly reared for meat, milk and skin. Goat meat is preferred meat in India. A goat on hoof (live goat) fetches a better price than a sheep on hoof.

Goats can be maintained under stall fed conditions. Successful goat rearing depends on the selection of site. Goats do not thrive on marshy or swampy ground. Goats are to be provided with a dry, comfortable safe and secure place, free from worms and affording protection from excessive heat and inclement weather.

Goat matures in about 6-7 months. Gestation period is 145-155 days. It gives birth to 1-3 kids per time. When the young ones attain a body weight of about 25-30 kg in about nine months, they can be sold.

The per head requirement of nutrients in respect of goats is relatively low. Hence they are suitable for resource poor farmers with marginal grazing lands.

9.3.9 Integrated sheep farming

Sheeps are well adapted in many areas. They are excellent gleaners and make use of much of the waste feed. They consume large quantities of roughage, converting a relatively cheap food into a good cash product. Housing need not be elaborate or expensive. However, to protect the sheep from predatory animals, the height of the fencing should be raised to 2 m.

There are three types of sheep in India based on the geographical division of the country.

1. *The temperate Himalayan region:* Gurez, Karanah, Bhakarwal, Gaddi and Rampur-Bushiar
2. *Dry western region:* Lohi, Bikaneri, Marwari, Kutchi and Kathiawari
3. *Southern region:* Deccani, Nellore, Bellary, Mandya and Bandur

One ram can be maintained for 40-50 ewes. Rams are liable to fight when two or more of them are put in a pen. Unlike other farm animals, ewes in general do not come in heat at regular intervals throughout the year but are seasonal in this respect. The duration of the heat period will range from 1-3 days and 75% of the ewes remain in the heat for 21-39 hours. The optimum time of service is towards the end of the heat period. Average heat interval is 18 days during the breeding season. The gestation period will vary from 142-152 days with an average of 147 days. A normal ram is in full vigor for breeding from the age of 2½ - 5 years. Sheep grow fully at two years of age when the ewe is ready for breeding. Under average range condition ewes may be expected to produce about five crops of lambs.

A sheep requires about 1-2 kg of leguminous hay per day depending on the age of the sheep and its body weight. Proteins may be supplied through concentrates such as groundnut cake, sesame cake or safflower cake, etc. Normally 110-225 g cake is sufficient to maintain an average sheep in good condition. Feeding a mixture of common salt, ground limestone and sterilized bone meal in equal parts is required to alleviate deficiency of minerals in the feed.

9.3.10 Integrated cattle rearing

Cattle and buffalo production is an integral part of rural economy and contributes substantially to the family income. Milk produces 63 per cent of animal protein and almost 100 per cent of animal fat in the daily diet of an average Indian. Milk contributed 66.8 per cent of the total value of the output from livestock (1998-99). In addition to milk and milk products for human consumption, cattle and buffaloes also provide animal power for agricultural operations and rural transport needs. The draught animal power, which is valued of Rs.4000-95000 million is not included in the total value of output from livestock. 75 million draught animals (mostly cattle and buffaloes) contribute 20 per cent of energy input into crop farming. Although there has been large reduction in contribution of draught from 72 per cent in 1961 to 23 per cent in 1991 mainly due to mechanization, the requirement of draught shall continue to be around 20 per cent in years to come. Around 54 per cent of this total milk production comes from buffaloes, 42 per cent from cows and 4 per cent comes from goats.

Cattle rearing in India is carried out under a variety of adverse climatic and environmental conditions. There are 28 breeds of cattle and 8 breeds of buffalo. The cattle are broadly classified into three groups

1. *Draught breeds:* The bullocks of these breeds are good draught animals, but the cows are poor milkers, e.g. Nagore, Hallikar, Kangeyam, Mali

2. *Dairy breeds:* The cows are high milk yielders but the bullocks are of poor draft quality, e.g. Sahiwal, Sindhi, Gir
3. *Dual purpose:* The cows are fairy good milkers and the bullocks are the good draft work capacity, e.g. Hariania, Ongole, Kankrej
4. *Exotic breeds:* The exotic breeds are high milk yielders, e.g. Jersey, Holstein-Freisian, Aryshire, Brown Swiss and Guernsey.

Important breeds of buffalo are Murrah, Niliravi (which has home tract in Pakistan), Mehsana, Suti, Zafarabadi, Godavari and Badavari. Of these Godavari has been evolved through crossing local buffaloes in coastal regions of Andhra Pradesh with Murrah.

Each cow requires 12 to 18 sq.m space and the buffaloes need 12 to 15 sq.m. It is important to provide good ventillation and an open shed of housing is always preferable. Dairy building should be located at an elevated place to facilitate easy drainage. The floor should be rough and have gradient of 2.5 cm for every 25 cm length.

9.3.11 Integrated fish farming

In many areas, the relative abundance of rainfall and the flat topography allows for the integration of fish production during part of the cropping cycle. The wide variation in the rainfall patterns of the lowland leads to a great diversity in integrated farming system. This diversity is also due to contrasting size of farms and tenure arrangements, proximity to markets and access to credit, infrastructure and inputs. The different uses of scarce land resources, the wide crop and animal diversity and many resource flows within the farm and beyond show the farmer's skill and creativity to make the most of this diversity.

Considering India's estimated population growth of more than one billion by the dawning of the third millennium, the total fish requirement will be more than seven million tonnes at the modest rate of 12.5 kg/capita/annum consumption. This is due to increasing trend in the fish eating habits of Indians from 27.7 per cent in 1987-88 to 39.7 per cent in 1996-97. India is also one of the major exporter of fisheries products and it constitutes about 4 per cent of India's total export earnings. India depends heavily on capture fisheries for fish production. The present fisheries production from both marine and inland resources is about 4.9 million tonnes, which is not sufficient to meet the demand. Furthermore, owing to a substantial increase in the fishing effort, a stagnation stage has been reached due to relentless unplanned exploitation of fisheries resource.

In integrated aquaculture system, animal manure applied into the ponds is first decomposed by bacteria and then nutrients including NPK. They are released into the water for the production of phytoplankton through photosynthesis. Phytoplankton is mainly grazed by the zooplankton, in turn, zooplankton is fed by benthos and so on, hence it forms a food chain. On the other hand bacteria are propagated at a high speed through the utilization of carbon in animal manure to form another food chain. These two food chains develop and overlap each other to set up a food web for various species of fish with different feeding habits.

9.3.12 Integrated biogas generation

Biogas is a popularly known renewable source of energy. Biogas production is based on the phenomenon of biological decomposition of organic materials in the absence of air. Biogas technology is however, some thing more than the mechanism of producing this useful fuel gas. Through this technology very good organic manure, improved rural sanitation and health care and more importantly prevention of deforestation and thus ecological balance can be achieved. Biogas, being one of the potential renewable sources of energy has a definite role to play because it reconciles two apparently conflicting objectives of getting cheaper and better fuel for cooking, lighting, running engines and good organic manure to supplement and optimize the use of chemical fertilizers from cattle dung and agricultural residues. Hence, a biogas plant is an asset to a farming community.

Integrated farming system may comprise a biogas plant, energy farming, windmills, solar heating and photovoltaic systems, food, fodder and algal cultivation and pisciculture. Effluent from the biogas plant provides nutritious protein-rich feed for animals particularly for fish. Solids are recovered from sludge only allowing it to settle in tanks or by draining and drying. These solids along with other additives after being meshed and detoxified can be used as animal feed. After separation of solids from sludge, remaining liquid can be utilized as a nutrient for algal growth. A large portion of biogas plant sludge is used as fertilizer for energy crops.

9.3.13 Integrated apiculture

Bee keeping is one of the most important agro-based industries. Nectar and pollen from flowers are the raw materials, which are available in plenty in nature. Bee keeping can even be started with a single colony.

There are two species, *Apis cerana* and *Apis mellifera* are complementary to each other but have different adaptations. *Apis cerana* is better acclimatized to higher altitudes of the Himalayan region. *Apis mellifera* is more profitable in the plains.

The most important sources of nectar and pollen are maize, great millet, sunflower and palm. The hive consists of floorboard, brood chamber, top cover, inner cover, frames and entrance rod. These parts can easily be separated. The hive may be double walled or single walled. The single walled hive is light and cheap. The most suitable time for commencing bee keeping in a locality is the arrival of the swarming season.

Honey should have good quality to meet the national and international standards. Qualities such as aroma, colour, consistency and floral sources are important. Proper honey straining and processing is needed to improve the quality of the produce. Honey varies in the proportion of its constituents owing to the differences in the nectar produced by different plants. The nectar collected by bees is processed and placed in comb cells for ripening. During the ripening, sucrose is converted into glucose and fructose by an enzyme called invertase, which is added to it by the bees. Honey is excellent energy food with an average of about 3500 calories per kg. It is directly absorbed into the human blood stream, requiring no digestion.

9.3.14 Integrated mushroom cultivation

Mushroom is an edible fungi with great diversity in shape, size and colour. Essential mushroom is a vegetable that is cultivated in the protected farms in a highly sanitized atmosphere. Just like other vegetables, mushroom contains 90% moisture with high in quality protein. Mushrooms are fairly good sources of vitamin C and B complex. The protein has 60-70% digestibility and contains all essential amino acids. It is also rich source of minerals like Ca, P, K and Cu.They contain less of fat and are considered good for diabetic and blood pressure patients.

a) Varieties

1. Oyster mushroom: *Pleurotus* sp. (var. CO1, CO2)
2. Paddy straw mushroom: *Volvariella volvacea*
3. White button mushroom: *Agaricus bisporus* (Var. A-11, Horst U3)

9.4 Integrated farming system for different regions

The overall objective of IFS is to evolve technically feasible and economically viable farming system models by integrating cropping with allied enterprises for different regions with a view to generate more income and employment from the farm.

9.4.1 Integrated farming system for low / wetlands

Cropping in low lands is considered to be less risky due to abundant availability of water. In addition, most of the low land soils are heavy clay type of soils, which are fairly fertile soil. Mostly rice is the principal crop in wetlands. Crops like banana, sugarcane and coconut are also grown in the ecosystems.

IFS can be adopted in the low lands with the following components such as fish, poultry, duck, mushroom, etc.

E.g. Rice – Fish – Azolla

Rice – Fish – Poultry / Duck

Fish – Duck

Rice – Fish – Poultry – Mushroom

Rice – Fish – Pigeon – Mushroom

Rice – Trees

Trees : For timber: *Tectona grandis*

For fruit: *Cocus mucifera*

9.4.1.1 Rice cum Fish culture

Due to water logging, the wetlands of rice fields serve as natural habitat as well as breeding ground for various kinds of fishes. In fresh water areas, there has been successful demonstration on raising Indian major carps viz., Rohu, Catla and Mirgal in perimeter trenches in rice plots, with Jaladhi from June to December and then Ratna and Jaya for four months. Production of deep water rice ranged from 800-1200 kg/ha/year.

In brackish water areas, rice fields are traditionally being utilized for culture of prawns and fishes in West Bengal and Kerala. In West Bengal, during monsoon in June or July, rice fields are manured and seedlings are planted. In August, bunds surrounding the fields are cut at places allowing the fish to enter the fields. They grow in fields during rice cultivation period and are harvested before harvesting of rice.

9.4.1.2 Grass -fodder cum fish culture

All or most parts of the crops planted in the fodder crop field and corner plots on the pond dykes and slopes are used as green fodder for the fish and as fertilizer for the ponds. The average yield of leguminous plants reaches up to 150 tonnes / ha or even more. Napier and berseem are more suitable for integration.

9.4.1.3 Fish cum duck integration

Annual manure production from ducks is 45-55 kg excreta/duck/year, which besides fertilizing the fish ponds also gets utilized directly as fish food. Apart from this 10-20 per cent of feed/day/duck is wasted, which are utilized in ponds. However, ducks do prey upon small fingerlings. Therefore, fish duck integration is suitable jointly in rearing and stocking ponds where fishes are above 12 g. In such integration of fish, species selection should be 60 per cent plankton feeder and 40 per cent omnivorous fish. Ducks are stocked @ 200-300 ducklings/ha of fishpond. In such integration, a production of 3500-4000 kg of fish, 18000-18500 eggs and 500-600 kg duck meat from one ha of pond area in one year without any supplementary feed and fertilizer can be obtained and the cost is turned down to 60 per cent than normal.

9.4.1.4 Aquaculture cum cattle integration

The total animal waste of cattle farming alone in India is estimated at 3000-4000 million tonnes annually. Among all the excreta from livestock, the amount of excreta of cattle is the greatest and most stable. Further cattle-cum-fish integration does not pose any serious prblem. The BOD of cattle manure is lower than that of any other livestock manure. It can be dispersed by hand, saves money, labour and energy and improves the environment. Therefore, aquaculture using cattle manure is safer than aquaculture with any other animal manure.

9.4.2 Integrated farming system for garden lands

The possibilities of having viable IFS in irrigated uplands are possible. The control and management of available resources in more efficient manner paves way to integrated upland compared to lowlands and rainfed lands. Components like dairy, poultry, goat, sheep, piggery, sericulture, mushroom, apiary, pigeon, rabbit, quail, etc. can be easily integrated in an integrated upland farm. In addition perennial trees like coconut and other fodder and multipurpose trees can be grown along the borders of the fields and boundaries of the farm.

e.g. Crop – Dairy – Biogas unit
Crop – Poultry – Biogas unit
Crop – Sheep / Goat rearing – Biogas unit
Crop – Sericulture
Crop – Piggery
Crop – Forest trees
Crop – Fruit trees
Crop – Trees – Cattle

Suitable timber tree species: *Tectona grandis, Dalbergia latifolia, Dalbergia sissoo, Ailanthus excelsa, Hardwickia binata,* etc.

Suitable fodder yielding tree species: *Leucaena leucocephala, Gliricidia maculata, Hardwickia binata,* etc.

Suitable fruit yielding tree species: *Cocus nucifera, Mangifera indica, Carica papaya, Psidium gujava, Citrus* spp., *Emblica officinalis, Ferronia elephantum,* etc.

Crops : Paddy, sorghum, maize, bajra, leguminous crops, etc.

9.4.2.1 Fish cum horticulture integration

Of all the system of integrated fish farming, the fish-horticulture integration appeals to the Indian farmers the most. In a year, three crops of vegetables are grown viz., summer - rainy season (kharif) and winter season (rabi). Commonly grown vegetables are bitter gourd, lady's finger, bottle gourd, brinjal, tomato, cauliflower, cabbage, potato, peas, palak, etc. Among fruit crops banana, papaya and coconut are considered the best. Pond bunds are also used for growing crops of pulses and oilseeds and also flowers. Fish-horticulture integration, on an average, adds to the income of fish farmers by 25-35% or even more.

9.4.2.2 Fish cum poultry integration

About 500-600 birds produce enough manure for one ha pond under polyculture. Poultry birds are kept on pens providing about 0.3 m^2 - 0.4 m^2 space/bird. Under this system litter from the poultry pens or fresh poultry droppings from the cage are recycled in the fish ponds which results in production of 4500-5000 kg fish, more than 70,000 eggs and about 1250 kg chicken meat from one hectare of pond area in one year, without any supplementary feeding or fertilizer.

9.4.3 Integrated farming system for drylands

The dry land ecosystem of Tamil Nadu is characterized by i) inadequate and uneven distribution of rainfall, ii)poor and marginal soils, iii)low cropping intensity, iv)limited crop diversification, v)low value crops and vi)poor resource mobilizing power of farmers.

The agriculture in drylands is seasonal. The cropping season is restricted to 4-5 months and people remain without employment for rest of the year. Diversification of cropping by integrating with components like livestock (sheep / goat rearing), silviculture, horticulture tree crops and pastures would improve the standard of living and employment opportunities of the dryland farmers.

IFS are a boon to dryland farmers. When compatible components / allied activities / associated enterprises are combined, farming in drylands become less risky and remunerative. IFS on a watershed approach are the best way of alternative in dryland areas. The following components are identified as appropriate to different situations such as goat / sheep rearing, agroforestry, farm forestry, horticultural tree crops.

e.g. Crop – Trees – Goat / Sheep / Cattle rearing

Crop – Trees (Fruit / Fodder / Timber)

Crop – Trees – Dairy

Crop – Trees – Dairy – Biogas

The suitable trees for fodder are *Leucaena leucocephala, Albizia lebbeck, Gmelina arborea, Gliricidia sepium, Acacia leucophloea, Acacia nilotica,* etc.

For fruits: *Ziziphus* spp., *Phoenix* sp., *Simarouba glauca, Emblica officinalis,* etc.

9.4.4 Economic analysis of IFS

Economic analysis of individual components revealed that gross return was higher in intercropping with tree species as compared to sole sorghum. The gross returns from the intercropping system were higher during first year than the second year. With regard to net return, the highest net return was recorded under cropping with *Emblica officinalis* (Rs.2096). The gross and net returns for goat component were Rs.13111 and Rs.995 and Rs.12527 and Rs.2123 for the first and second year, respectively.

Cropping with *Emblica officinalis* or *Ailanthus excelsa* with goat recorded the highest B:C ratio of 1.18 during first year. During second year, the highest gross and net return and B:C ratio (Rs.16536, Rs.1143 and 1.07 respectively) were recorded in cropping with *Emblica officinalis* with goat component.

Farming with sole sorghum provided employment for 37 man days/ha/yr, whereas the cropping systems in IFS generated 44 man days /ha/yr during first year and 40 man days in the second year. Integration of trees generated additional employment opportunity for 17 man days /ha/ yr. Inclusion of goat component generated employment opportunity for 23 man days during first year and 35 man days during second year. Additional employment of 47 and 55 man days during first and second year generated through IFS provided opportunity for the employment of family labour who are idle during the lean season. Of the total employment generated by the IFS, crop contributed 52.4% and the contribution of trees and goat were 20.2 and 27.4 per cent, respectively in the first year. During second year, the contribution of crop, tree and goat components were 43.5, 18.5 and 38.0 per cent respectively.

Livestock is the best complementary enterprise with cropping, especially during the adverse years. Installation of biogas plant in crop-livestock system will make use of the wastes, at the same time provides valuable manure and gas for cooking and lighting. In a wetland farm, there are greater avenues for fishery, duck farming and buffalo rearing. Utilizing the rice straw, mushroom production can be started. Under irrigated conditions (garden lands), inclusion of sericulture, poultry, piggery along with arable crop production is an accepted practice. The poultry component in this system can make use of the grains produced in the farm as feed. Pigs are the unique components that can be reared with the wastes, which are unfit for human consumption. In rainfed farming, sheep and goat rearing form an integral part of the landscape.

Agroforestry is the activity which can be included under dry land conditions. In this system, selection of enterprise should be on the cardinal principle that there should be minimal competition and maximum complementary effect among the enterprises. Integrated farming system models are region specific. Hence to improve the productivity from the existing farming system integrated farming system research is needed to meet the demand in present and future.

Questions

1. Judicial management of resources in such a way to sustain productivity and high environment quality is termed as

   a) Integrated farming system　　b) Integrated nutrient management

   c) Integrated Resource management　　d) None of the above

2. Phytoplanktons and zooplankton are major food for

   a) Fishes　　b) Prawns

   c) Crabs　　d) All the above

3. Fish production can be increased through

   a) Poly culture　　b) Mono culture

   c) Both　　d) None of the above

4. High BOD in the pond is the indication of

   a) Purity of water b) Oxygen demand

   c) Microbial status d) None of the above

5. Presence of excess growth of algae and nutrient richness in ponds leads to a phenomenon called as

   a) Eutrophication b) Algal bloom

   c) Algal culture d) Biomagnification

6. The main host plant of silk worm is

   a) Mulberry b) *Schleichera oleosa*

   c) *Zizyphus* sp d) *Acacia melanoxylon*

7. The protein rich food preferred by fishes

   a) Silk worm pupae b) Silk worm larvae

   c) Both d) None

References

Jayanthi, C., A. Rangasamy and C. Chinnasamy. 1997. Economic efficiency of component linkage in low land integrated farming systems. Madras Agricultural Journal, 84(8): 498-504.

Palaniappan, S.P. and K. Sivaraman. 1996. Cropping system in the tropics. Principles and management (2nd Edn.). New Age International Pvt. Ltd., New Delhi.

Singh, S.S. 1997. Crop management under irrigated and rainfed conditions. In: Principles of Agronomy. Kalyani Publishers, Ludhiana, India.

Venkataraman, V., S. Subramanian, R. Kulandaivelu, R. Srinivasan, N. Balasubramanian and V. Veerabadran. 1983. Integrated farming system - An experience and case studies. TNAU, Coimbatore.

—•—

CHAPTER **10**

Agroforestry and Watershed Management

Tree species are considered important component of agriculture especially in a catchment area. The vegetative cover to the ground surface reduces the runoff considerably and promotes the maximum rain water infiltration into the ground. In agroforestry system, roots of trees and crops cover a large area under the soil surface, tree-crop root system losses the soil, which increase the infiltration rate. In watershed area a large hilly terrain which are only partially cultivated can be made to yield high returns per unit area by judicious combination of suitable tree species and crops/ legumes/ fodder along with suitable water management techniques.

10.1 Different Terminologies

a) Watershed

Watershed is a geohydrological unit or a place of land that drains at a common point (or) land area from which water drains to a common point. In this content, watershed is considered to be synonymous with 'catchment' and 'drainage basin'.

According to Mohsin, the total area contributing water, which flows through the outlet is the watershed of that outlet. The outlet may be a small natural drain, a rivulet, a stream or even a river.

A tertiary stream or river is a tributary of secondary river having a drainage area of 4,000 to 40,000 ha. The drainage area of such a stream river is called a watershed.

b) Catchment

A primary river is a long river having a drainage area of more than one lakh hectare with a number of tributaries of first, second, third, fourth or even fifth order. The primary river generally discharges load to the sea. The drainage area of such a river is called a catchment.

c) Sub-catchment

A secondary river is a tributary of primary river having a drainage area of more than forty thousand hectare. The drainage area of such river is called a sub-catchment.

d) Sub-watershed

A quaternary stream is a tributary of a terrain having a drainage area between 2,000 to 4,000 ha.

e) Mini-watershed

A pentad streamlet is a tributary of a quaternary having drainage area between 400-2000 ha. Such a drainage area is called mini-watershed.

f) Micro-watershed

A hexed streamlet is a tributary to a pentad having a drainage area of less than 400 ha. Such a drainage area is called micro-watershed.1

10.2 *Objectives of watershed management*

As per NWDPRA guidelines, the objectives are as follows.

i) Conservation, upgradation and utilization of natural endowments like lands, water, plant, animal and human resources in a harmonious and integrated manner. This will aim at perpetual availability of food, fodder, fuel, fibre, timber and biomass for rural and cottage industries to meet the growing demands of human and livestock population through diversified land use.

ii) Generation of massive employment during the project period to the rural population like small and marginal farmers and landless labourers, tribals, etc.

iii) Improvement of production, environment and restoration of ecological balance through scientific management of land and rain water

iv) Reduction of inequalities between irrigated and rainfed areas. Ultimately, stable production and processing of biomass would contribute towards better life in rural areas. This will reduce large scale migration from rural areas to cities.

v) In addition to food, fuel and fodder the project would endeavour to enhance cash flow to the rainfed farmers and landless agricultural labourers through increased casual employment, marketable surplus of agricultural and dairy produce, growing of cash crops like vegetables, etc. in suitable areas.

10.3 *Role of hydrological cycle in watershed management*

Hydrological cycle defined as the continuous circulation of water among the hydrosphere, atmosphere and lithosphere. The physical process controlling the distribution and movement of water can be understood in terms of the hydrological cycle. There is no beginning and ending point in the hydrological cycle. Moisture is constantly rotating from the ocean to the atmosphere, to the land and back to the ocean.

As the population increases, changes to the land have been significant, with major changes in the runoff characteristics of watershed. Land clearing for agriculture increases the amount of exposed soil with a corresponding decrease in the protective cover of the natural vegetation. Consequently, it results in reduction in infiltration, increased surface runoff and tremendous soil losses. The reduced storage due to heavy sedimentation results in increased rate of runoff which increases both flood damage and channel erosion. Thus, failure to understand the process of the hydrologic cycle may result in losses to life and property.

Knowledge of hydrology helps in balancing the demand for water with supply to prevent floods and to protect the quality of streams and lakes, with the perspective of the watershed and a clear awareness of the linkages between upland and downstream areas. It is important to plan and develop long-term sustainable solutions to many natural resources problems and at the same time, avoid different types of environment degradation. The hydrological cycle is complex and dynamic but can be simplified on the principle of conservation of mass in terms of input, output and storage. The hydrological processes and the effects of vegetation and soils on these processes are of particular interest in watershed planning and management.

10.4 Biological measures for Watershed management

The common biological measures to tackle the problems of watershed areas are as follows.

a) Choice of cropping system

b) Suitable crops, varieties, sowing time, seed rate and use of fertilizer

10.4.1 Choice of cropping systems

The following agroforestry systems are recommended for watershed areas.

a) Alley cropping

The system involves planting of rows of trees or shrubs along the contour and food crops and cover crops are grown in between the rows. In this system, three types viz., i) Intensive alley cropping system, ii) Alley cropping with perennial grass in watershed area and iii) Wide alley cropping system can be adopted. Its main objectives are i) to conserve moisture and soil, ii) to improve nutrient status and other properties of soil and iii) to increase productivity by optimum utilization of the land.

In Intensive alley cropping system, the following species viz., *Leucaena leucocephala,* perennial pea and vetiver grass can be grown intensively for effective watershed management.

In Alley cropping with perennial grass in watershed area, the cropping systems viz., i) Hedge forming species + grass species + FYM, ii) Hedge forming species + grass species + nitrogen application and iii)Hedge forming species + grass + tree foliage as green manure can be followed.

In Wide alley cropping system, the following trees species viz., *Acacia nilotica*, *Azadirachta indica,* Eucalyptus, *Sesbania grandiflora, Zizyphus* spp. and the following agricultural crops viz., sorghum, pearl millet can be grown as intercrops effectively.

b) Silvipasture system

In silvipasture system some of the following fodder trees are recommended viz., *Bauhinia variegata, Toona ciliata, Albizia amara, Acacia tortilis, Morus alba, Grewia oppositifolia, Gmelina arborea, Leucaena leucocephala* and the following grasses viz., *Bamboo* spp., *Saccharum munja, S. spontaneum, Panicum* sp. are recommended.

c) Agrosilvipastoral system

In this system, the following tree species viz., *Acacia catechu* or any nitrogen fixing trees along with any of the following crops viz., pearl millet, sorghum, sesame, castor, mung bean, cowpea and any of the grasses viz., *Saccharum bengalensis, Panicum antidotale, Cenchrus ciliaris, C. setigerus* and *Lasiurus indicus* are recommended.

10.4.2 Suitable crops and cultivars, sowing time and use of fertilizer

a) Suitable crops and cultivars

The undulating topography and soil with low moisture retentive capacity restrict the choice of crops or its cultivars in the watershed area. For dryland areas, the crop variety should be drought resistance, short duration and should also be early vegetative cover (e.g. safflower); deep root system (safflower, Bengal gram) and drought and disease resistance (lentil, barley).

b) Sowing time

For Kharif crops, land preparation is to be done at pre monsoon showers and sowing should be started soon after receiving the monsoon.

c) Nutrients

In watershed area under agroforestry system, best method to improve the nutrient status of soil is the use of green manures and dung manures. The use of leguminous species increases the organic contents of the soil which improves the water holding capacity of the soil.

Questions

1. The hydrological unit that drains at a common point is called

   a) Water shed b) Catchment
   c) Drainage basin d) All the above

2. The drainage area of tertiary stream or river is

   a) 4,000 to 40,000 ha b) 4,400 to 45,000 ha
   c) 4,500 to 50,000 ha d) 1,000 to 15,000 ha

3. The drainage area of catchments is

   a) More than one lakh hectare b) More than forty thousand hectare
   c) More than seventy five thousand d) None of the above

4. The drainage area of a sub-catchment is

   a) < 1 lakh ha b) < 40,000 ha
   c) 2, 000 – 4,000 ha d) 400-200 ha

5. The continuous circulation of water among the hydrosphere, atmosphere and lithosphere is called as

   a) Nutrient cycle b) Hydrological cycle
   c) Water cycle d) None of the above

6. The drainage area of sub water shed is

   a) 1000-1500 ha b) 2000-4000 ha
   c) 5,000 ha d) <5,000 ha

7. The drainage area of a mini watershed is

   a) <2000 ha b) 2000-3000 ha

   c) 400-2000 ha d) 2500 ha

8. The drainage area of a micro water shed is

   a) < 400 ha b) > 400 ha

   c) < 1000 ha d) 400-1000 ha

References

Gupta, J.P. and B.M. Sharma. 1997. Agroforestry for sustained productivity in arid region. Scientific Publishers, Jodhpur, India. 198 p.

Jha, L.K. 1995. Advances in agroforestry. APH Publishing Corporation, New Delhi. 669 p.

Young, A. 1989. Agroforestry for soil conservation. CAB International and ICRAF, UK.

—•—

CHAPTER **11**

Role of Trees in Soil and Water Conservation and Soil Fertility

11.1 Role of Trees in Soil and Water Conservation

The establishment of a vegetative cover is one of the most effective means of soil and water conservation. When the protective cover of vegetation on the soil is removed, the structurally unstable tropical soils are exposed to the beating action of rains. Losses due to erosion immediately after land clearing are normally alarmingly large. The potential role of trees in reducing run-off and erosion losses is well appreciated and understood. Natural forest communities provide a multilayer defence against the impact of raindrops. The different state of canopy progressively reduces the force of rain, thereby reducing the adverse effect of its impact on the soil. Further more the litter and the humus layers on the soil surface act as a cushion against erosion.

The soil conservation is particularly important in highlands with undulating topography and steep slopes, which are increasingly being brought under cultivation. For example, in South-East Asia, there is a long tradition of planting *Leucaena leucocephala* in contour hedgerows for erosion control and soil improvement. These countour rows of *leucaena* survive throughout the long dry season because of their long tap roots, which can reach water deep in the ground. Loppings and prunings from the hedgerow species also provide mulch to aid in preventing sheet erosion between trees. Removal of vegetative cover from the soil generally results in an increase in bulk density, a decrease in porosity and reduction in infiltration rate.

The use of trees and other woody perennials to protect agricultural fields from these adverse effects is a wide spread practice. Trees acting as windbreaks and shelter belts also assist in regulating the microclimate within the tree stands. These also reduce evaporation and temperature. The clearing of vegetation affects not only the farmlands in the immediate vicinity, but also destroys the water catchment areas causing flooding of rivers and rapid silting of dams.

11.1.1. Soil Conservation Methods

The following are the different vegetative methods used for soil and water conservation.

a) *Strip planting:* In this method, erosion permitting and erosion resisting crops are alternatively raised at right angle to the slope of the land to retard the velocity of rain water

b) *Rotational cropping:* In this method, either grain crops grasses or legumes along with trees are planted in the field. This will help to improve and maintain soil fertility.

c) *Cover cropping:* Trees and grasses are grown to cover the earth's surface. Trees like *Acacia nilotica, Azadirachta indica, Eucalyptus tereticornis,* silk cotton, teak and casuarina can be planted to arrest erosion along with trees. *Agave americana* can be planted for stabilization of gullies. *Acacia nilotica and Azadirachta indica* can be planted on the banks of rivers, percolation ponds, lakes to strengthen the bunds and to prevent erosion.

11.2 Role of trees in soil fertility

Soil is one of the most important natural resources to suffer as a result of tree cutting. If it is not protected, its productivity declines and it may become difficult to sustain the human and animal population even at its present level. Therefore, protection of this resource is important and an understanding of how this resource is influenced in an agroforestry system is necessary.

It is generally assumed that trees have deep and spreading roots and hence are capable of exploiting more soil volume and taking up nutrients and water from deeper layer not usually contacted by herbaceous crops. This process of taking up nutrients from deeper soil profiles and eventually depositing at least some portion of them on the surface layer through litter-fall and other mechanisms is referred to as *'nutrient pumping'* by trees. It is well known that the development of plants depends on site characters and environmental factors. Many woody species have the largest number of roots and the majority of the fine roots are located in the uppermost fertile portion of the soil profile. Some tree species are shallow rooted. *Prosopis chilensis* has a shallow and spreading root system whereas *Prosopis juliflora,* is known to have a very deep root system.

11.2.1 Soil fertility Improvement Processes

The tree improves the soil fertility by the following processes.

a) Organic matter and nutrient addition to the soil

Tree species contain large quantities of 'living' biomass .About 20 to 25 per cent of the total living biomass of the trees is in roots and there is a constant addition of organic matter to the soil through dead and decaying roots. The major addition of organic matter and other nutrients to the soil from the trees standing on it is through litter fall i.e., dead and falling leaves, twigs, branches, fruits etc.

b) Dinitrogen fixation by trees

There is a possibility for improvement of the fertility status of agricultural lands through additional amounts of nitrogen added to the soil by the tree legume component. Mimosoideae and Fabaceae are well known to fix nitrogen. Therefore, among the various avenues of addition of nitrogen to a soil through natural and biological means, the most significant one brought about by on agricultural lands nitrogen fixation by leguminous trees. For example, *Leucaena leucocephala* grown for forage for 9 months yields about 12,600 kg forage, 3,600 kg protein and 575 kg nitrogen per ha.

c) Nutrient cycling

The nutrient cycling model consists of the soil-plant system that is partitioned into several compartments. The crown surface forms the boundary of the system where input of bioelements occurs through precipitation. The soil surface is the entry point for inputs into the soil compartment, occurring through fertilizers, rainfall and stem flow. The surface layer may be considered the zone of intensive root activity, with the subsoil constituting the extensive root activity zone. The lower end of the extensive root layer is the boundary of the ecosystem to the hydrosphere and lithosphere. Bioelements transported beyond this layer are lost from the ecosystem and appear as output from the system.

Questions

1. The practice of rotational cropping helps in

   a) Improving and maintaining soil fertility
   b) Soil and water conservation
   c) Both
   d) None

2. Erosion permitting and erosion resisting crops grown alternatively at right angle to the slope of the land is called as

   a) Strip planting
   b) Linear planting
   c) Cover cropping
   d) None of the above

3. The process of taking up nutrients from deeper soil profiles and depositing them on the surface layer is referred to

   a) Nutrient uptake
   b) Nutrient absorption
   c) Nutrient pumping
   d) None of the above

4. The process of cycling of nutrients from soil to the plant and back to the soil is called as

   a) Nutrient cycling
   b) Nutrient enrichment
   c) Nitrogen pumping
   d) Bio geo-chemical cycling

5. Soil fertility is improved by

   a) Addition of organic matter to the soil
   b) Nitrogen fixation by trees
   c) Nutrient cycling
   d) All the above

6. The process by which the complex organic substances are broken down to simple inorganic elements is called as

   a) Mineralization b) Nitrification

   c) Metabolism d) None of the above

References

Bandyopadhyay. 1997. A textbook of agroforestry with application. Vikas Publishing House Pvt. Ltd. 117 p.

Gupta, J.P. and B.M. Sharma. 1997. Agroforestry for sustained productivity in arid region. Scientific Publishers, Jodhpur, India. 198 p.

Young, A. 1989. Agroforestry for soil conservation. CAB International and ICRAF, UK.

—•—

CHAPTER 12

Wastelands Development

Wasteland is that land which is presently lying unused or which is not being used to its optimum potential due to some constraints.

12.1 Classification

National wastelands development board classifies wastelands into two major categories as below.

1. Cultivable wastelands
2. Uncultivable wastelands

The cultivable wastelands are classified into:

1. Gullied and/or ravinous lands
2. Undulating land with without scrub
3. Water logging and marshy lands
4. Salt affected lands
5. Shifting cultivation area
6. Degraded forest lands

7. Degraded pasture / grazing land
8. Degraded forest plantations
9. Sand dunes
10. Mining / industrial wastelands

Uncultivable wastelands which cannot be used for vegetation are classified as :

1. Barren rocky / stony area
2. Steep slopy areas
3. Snow covered and / or glacier lands

12.2 Extent of Wastelands

Wastelands statistics (2003) indicate that about 63.85 million hectares, which account for 20.17% of the total geographical area (328.72 million ha) exist as wastelands in India (Table 1).

12.3. Development of different wastelands through agroforestry

12.3.1 Gullied / Ravinous land

Ravines are deep gorges, comprising of a system of gullies, running more or less parallel to each other, in deep alluvium and draining into a major river or its tributary which is following much lower than the surrounding table lands. Ravines represent the last stage of water erosion. Some of the causes of ravine formation are as follows :

i) Clearance of forest for agriculture purpose
ii) Unscientific agricultural methods of soil working
iii) Burning of ground cover and uncontrolled grazing
iv) Low annual rainfall
v) Extreme dryness of soil due to easy drainage and high soil porosity

Table 1. Categorywise wastelands in India

Category of Wastelands	Area (sq. km)	% of total geographical area covered
Gullied and/ or ravinous land	20553.35	0.65
Land with or without scrub	194014.29	6.13
Waterlogged and marshy land	16568.45	0.52
Land affected by salinity/alkalinity-coastal/inland	20477.38	0.65
Shifting cultivation area	35142.20	1.11
Under-utilized/degraded notified forest land	140652.31	4.44
Degraded pastures/grazing land	25978.91	0.82
Degraded land under plantation crop	5828.09	0.18
Sands-Inland/coastal	50021.65	1.58
Mining/industrial wasteland	1252.13	0.04
Barren r ocky/stony waste/sheet rock area	64584.77	2.04
Steep sloping area	7656.29	0.24
Snow covered and/or glacial area	55788.49	1.76
Total wastelands	638518.31	20.17

(Wastelands Statistics, 2003)

These ravinous lands can be made positive through integrated approach of soil conservation and afforestation. Maintaining dense multistorey vegetation is considered highly effective. The efficiency of afforestation through agroforestry are as follows.

1. It helps to check slash erosion
2. It reduces runoff and aids infiltration

3. It improves soil structure through the addition of organic matter

The species suitable for agroforestry includes *Acacia nilotica, Azadirachta indica, Dalbergia sissoo, Tectona grandis, Albizia lebbeck, Leucaena* and *Eucalyptus* spp., etc.

Among grasses and legumes, *Chrysopogon fulvus, Cenchrus ciliaris, Pennisetum, Sacharum spontaneum, Dicanthium annulatum* and *Stylosanthes* spp. are found suitable for such lands.

Gullied lands could be rehabilitated by adopting the following measures.

i) Retention of runoff in the drainage area
ii) Diversion of runoff in the drainage area
iii) Conveyance of runoff through a gully

The gullied lands could be developed into silvipastures by using suitable plant species after proper protection from biotic interference by fencing. The grasses can be raised through direct seeding but for the tree seedlings raised in polyethylene bags should be used. The spacing between plant-to-plant and row to row should be 5 m.

12.3.2 Undulating Land with or without scrub

These lands are prone to degradation and erosion. They occupy high areas of topography but excludes mountainous terrain. It is mainly associated with plateau, laterite formations and bhata lands in the states of Maharashtra, Andhra Pradesh and Rajasthan. These areas are met with conditions of high rainfall and undulating terrain.

The causes for the formation of such lands include deforestation and its rehabilitation is possible only through agroforestry. It has been recommended to go for interplanting *Agave sisalana* with palatable grasses that has high efficiency to protect soil from erosion. The tree species suitable for these wastelands are *Eucalyptus* spp., *Dalbergia sissoo, Gmelina arborea, Emblica officinalis, Acacia auriculiformis, Anogeissus pendula, Albizia procera* and *Bamboo* sp.

12.3.3 Waterlogged and marshy lands

The main causes for the formation of these lands are canal seepage, high water table and improper drainage. It occurs in low lying areas where water stands for most of the year. These lands are reclaimed by establishing drainages like cut off drains, collector drains and evacuator drains. This is followed by going for afforestation with species that are tolerant to high water table and with high transpiration rate. The suitable species for these lands are *Eucalyptus robusta, Anthocephalus chinensis, Syzygium cumini, Salix* spp. and *Terminalia arjuna.* A combination of agricultural crops like jute, sugarcane with palatable grasses like Napier and Guinea grass is highly effective.

12.3.4 Land affected by salinity / alkalinity

These wastelands are found mainly in states of UP, Gujarat, Punjab, Haryana and Rajasthan. The factors responsible for the development of such wastelands are as follows

i) Arid and semi-arid climate
ii) Hard pan in sub-soil
iii) High water table
iv) Basin shaped topography
v) Impeded drainage conditions
vi) Excessive irrigation leading to seepage using saline water
vii) Ingress of sea water

These lands are classified into four classes viz., i) saline, ii) saline-alkali, iii) non-saline and iv) alkaline soil. These lands are rich in soluble salts and exchangeable sodium.The best remedy is going for afforestation by adopting different agroforestry systems. The suitable grass species for these wastelands are *Diplachure fusca, Chloris gyana* and *Brachiaria mutica.*

It is reported that, growing bermuda grass for two years coupled with flooding twice monthly with irrigation water proved to be as effective as the application of gypsum. The best trees suitable for these wastelands are *Prosopis juliflora, Eucalyptus* spp., *Dalbergia sissoo, Acacia nilotica, Albizia lebbeck* and *Azadirachta indica.* The effect of *Prosopis juliflora* is found good for the reclamation of saline - sodic soil.

12.3.5 Shifting cultivation

Shifting cultivation is a cyclic land use practice of felling trees and burning of forest areas for the purpose of growing agricultural crops leading to degradation. The practice is highly dominant in north-eastern states and major limitations for its development includes,

i) Shallow to moderate deep soil
ii) Irregular slope
iii) Presence of drainage channels
iv) Very severe water erosion
v) Constant human pressure

The best alternatives suitable for shifting cultivation are i) Agricultural land use, if slope is upto 50% with soil depth of >1m, ii) Horticultural land use, if slope is greater than 100% with soil depth of <1m, iii) Agri-horti-silvipasture, if slope is 100% with soil depth of >1m and iv) Livestock based land use, if slope is 100% with soil depth of <1m.

12.3.6 Sandy areas

The sand dunes and sandy undulating plains are responsible for the creation of such type of wasteland. The new sandy wastelands are being created due to increasing anthropogenic activities like cultivation, cutting and lopping of trees and over grazing. These wastelands are characterized by :

i) Loose structureless sands

ii) Poor fertility and water holding capacity

iii) Presence of alternate layers of $CaCO_3$

These lands are rehabilitated by adopting agroforestry practices since the land is rich in minerals and nitrogen and hence supports vegetation.

The best recommendation is planting *Casuarina* along the coasts. Other way is planting *Casuarina* in the outer area and growing cashew nut and coconut in the inner area. *Acacia tortilis* is found best on these sandy areas since the afforestation not only stabilize sand dunes but also ameliorate climate and vegetation of that area. The best species suitable for these sandy areas are *Casuarina equisetifolia, Ficus* spp., *Calophyllum inophyllum, Anacardium occidentale* and among grasses *Sorobolus* spp., *Cyanodon dactylon, Panicum maximum, Cenchrus* spp., etc.

12.3.7 Mining / Industrial wastelands

Mining is characterized by :

i) Undulating topography

ii) Presence of pits and debris of various dimensions due to mining

iii) Water and wind erosion due to irregular slopes

Reclaiming these lands is a challenging task due to drastically distributed ecosystem with poor physical, chemical and biological soil properties. The species found suitable with high survival rate are *Eleagnus umberilata, Coriaria nepalensis, Indigofera pulchella,* etc. For Bauxite mined areas, the species suitable are *Prosopis juliflora, Eucalyptus,*

Cassia siamea, Inga dulce, Acacia sp. and *Albizia lebbeck.* The following techniques found suitable to afforest mined out areas.

a) Covering toxic material with fertile soil

b) Neutralizing pH of soil using lime and gypsum

c) Leaching of excess salts by means of rain water

d) Stabilizing the surface with mulch

e) Silvipasture is found highly successful in establishing vegetation in mined out areas.

12.3.8 Under utilized / degraded notified forest land

These lands are characterized by eroded and undulating topography with very steep slopes and hence the water and wind erosion occurrence is very high. The growth of trees and shrubs is very poor with density being <20 per cent. Hence in order to restore its productivity, planting the following species viz., *Terminalia* spp., *A. nilotica, Agave, Albizia lebbeck, A. saman, Chloroxylon swietenia,* etc. is found highly effective

12.3.9 Degraded pastures / grazing land

This type of wasteland develops mainly on rocky topography, saline alluvial plains and eroded uplands and are characterized by stunted plant growth, irregular slopes, sandy and sandy loam soils.

The recommended remedial measures are establishing soil and water conservation measures like wind breaks, shelterbelts, mulching, contour bunding, etc. to upgrade soil productivity followed by establishing tree cover with species like *Acacia tortilis, Acacia senegal, A. nilotica, Butea* spp., *Panicum antidotale, Indigofera cordifolia, Heteropogon* spp., etc.

12.3.10 Snow covered / glacial areas

It is created mainly by glacial and fluvial glacial activities in mountainous area. These lands are covered by extensive snow and the recommended species for rehabilitation includes *Pinus gerardiana, Daplux acuminata, Desmodium* spp., *Butea minor, Salix* sp., etc.

Questions

1. NWDB classified wastelands into

   a) 2 types b) 3 types

   c) 4 types d) None of the above

2. Extend of wastelands in India

   a) 63.52 m ha b) 80.95 m ha

   c) 45.65 m ha d) None of the above

3. The land comprising deep gorges, gullies in deep alluvium and draining into a major river is known as

   a) Gullied/ ravinous land b) Eroded land

   c) Watershed d) Water logged area

4. The last stage of water erosion is the formation of

   a) Gullies b) Ravines

   c) Drill d) Sheet

5. The wastelands formed due to canal seepage and improper drainage is

   a) Gullies /ravinous lands b) Water logged and marshy lands

   c) Mangroves d) Sand dunes

6. The lands which are rich in soluble salts and exchangeable sodium

   a) Saline b) Alkaline

   c) Saline- alkaline d) All the above

7. The suitable species for reclamation of saline and sodic soils is

   a) *Prosopis juliflora* b) *Albizia lebbeck*

   c) *Azadirachta indica* d) *Casuarina equisetifolia*

8. The most suitable species for reclamation of sandy soils is

   a) *Casuarina equisetifolia* b) *Acacia tortilis*

   c) *Cynadon dactylon* d) *Panicum maximum*

References

Anil kumar and R.N. Pandey. Wasteland management in India. Ashish Publishing House, New Delhi. 227 p.

Gupta.R.K 1993. *Multipurpose Trees for Agroforestry and Wasteland Utilization*. Oxford and IBH Publishing company, NewDelhi. 580p

Singh, M.P. and D.N. Tewari .1996.Agroforestry and Wastelands. Anmol Publications Pvt. Ltd, New Delhi.229p.

—•—

CHAPTER 13

Agroforestry Diagnosis and Design

Agroforestry D & D is a family of procedures for the diagnosis of land management problems and potentials and the design of agroforestry solutions. The ICRAF has developed an approach to assist agroforestry researchers and development fieldworkers to plan and implement effective research and development projects.

13.1 Key Features of D & D

i) *Flexibility:* D & D is a flexible discovery of procedure, which can be adopted to fit the needs and resources of different users

ii) *Speed:* D & D has been designed with the option of a 'rapid appraisal' application at the planning stage of a project with indepth follow-up during project implementation.

iii) *Repetition:* D & D is an open-ended learning process. Since initial designs can almost always be improved. The D & D process need not end until further improvements are no longer necessary.

Basic logic of AF Diagnosis and Design

Basic question	Key factors to consider
Prediagnostic stage Which land-use system? How does the system work?	Distinctive combinations of resource technology and land user objectives. Production objectives and strategy subsystems and components.
Diagnostic stage How well does the system work?	Problems in meeting objectives, causal factors, constraints and intervention points.
Design and evaluation stage How can the system be improved?	Specifications for problem-solving or performance enhancing interventions.
Planning stage How can the improved technology be developed and disseminated?	D & D needs, extension needs.
Implementation stage How can the plan of action be adjusted to new information?	Feedback from research trials, independent farmer innovations, etc.

13.2 Procedures of AF Diagnosis and Design

The basic unit of D&D analysis is the land use system (LUS). The LUS can be defined and analyzed at the level of a country, ecozone, farming system, crop system, or any other unit. The structure and function of any LUS are determined by climatic, physical, biological, technological, economic, social and political factors. D&D focuses on the interactive effects these factors have on the LUS, and searches for opportunities for improved system development in the LUS.

D&D can be done at two levels i.e., Macro and Micro. Macro D&D is a large-scale analysis of an ecozone within a country or a group of countries. Macro D&D is important for deciding on national agroforestry research and extension agenda at the national level. Micro D&D, in contrast, focuses on one land use system (LUS) within the larger ecozone that has special priority for agroforestry intervention. Micro D&D involves a detailed analysis of households and production systems in the LUS.

13.2.1 Macro D&D

13.2.1.1 Objectives of Macro D&D

Macro D&D is an analysis of an ecozone within a country or group of countries. The four main objectives of the Macro D&D are as below.

i) To identify broad issues and problems constraining all the land use systems in a given ecozone

ii) To identify and prioritize areas for potential agroforestry interventions

iii) To identify research priorities and formulate research programme and

iv) To identify needs, opportunities, and mechanisms for inter-institutional collaboration for technology development.

To meet these objectives, macro D&D uses rapid appraisal techniques. It relies heavily on secondary data, which are verified and complemented by quick field surveys.

13.2.1.2 *Steps in Macro D&D Exercise*

There are seven steps in macro D&D exercise as below.

i) Identification of study ecozone

ii) Delineation of land use systems within the ecozone

iii) Description of land use system

iv) Analysis of land use system constraints and potentials

v) Analysis of potential agroforestry technologies

vi) Definition of agroforestry research needs and

vii) Inter- institutional coordination.

a) Identification of study ecozones

The first step in macro D&D is the selection of an ecozone for study. The zone covered in macro D&D exercise is large, containing significant variations in land characteristics with respect to current uses and constraints. The choice of the study zone should reflect its biological and socioeconomic importance at the national level based on the following factors.

a) The zone's contribution to food production and/ or income

b) The total population it supports and /or the area it covers

c) The urgency of its constraints

d) The extent of its unexploited potential for production and

e) The level of its development with respect to other areas.

b) Delineation of land use systems

The main guideline for distinguishing land use systems is that each system should display unique constraints and potentials differentiating it from other systems in the ecozone of interest.

Accordingly, a LUS consists of a distinctive combination of soils, crops, livestock, trees and/or other production systems; it occupies a given unit of land where specific outputs are desired and obtained by a given management unit. Normally the smallest unit of decision - making is the household, but any unit (i.e., communal group, cooperatives or company) that makes management decisions collectively and/or shares intimately in the input/output flows of a system is also considered to be a LUS unit.

c) Description of land use systems

All delineated LUSs are described by specifying the characteristics that are known to affect their current management and performance, and would be expected to affect the introduction of potential agroforestry technologies.

d) Analysis of land use systems constraints and potentials

Each system has to be evaluated for factors that prevent its households from obtaining optimal outputs from the available resources. This step requires analysis of farmers' need and priorities to see how well these are being met by current performance of the LUSs. The performance gap is evaluated by comparing the present levels of outputs with the biophysical and socioeconomic potential of the resources. For instance, one can compare the range of yields obtained in different LUSs with yields obtained in on-station or on-farm research. Constraints analysis is based on problems facing households both present problems and envisaged future problems. Emphasis is put on constraints which agroforestry can address. To diagnose constraints properly, the relationships between manifested symptoms and causal factors must be understood clearly.

Because every constraint identified may have several causal factors and symptoms (effects), the D&D team must have a multi-disciplinary capability. It must be able to interpret the relationships between these factors and the objectives of the household. Further more, it must be able to determine what opportunities exist to address the constraints. For this reason, constraint analysis is done concurrently with LUS characterization. For example, if one observes steep slopes in crop land, one can conclude that soil erosion is a likely hazard if nothing is being done to prevent it.

e) Analysis of potential agroforestry technologies

In this step, potential interventions are identified and assessed for their relevance to the constraints and their likelihood of increasing or sustaining productivity of the LUSs. First, all possible interventions are identified, whether they are from the areas of agronomy, forestry, or agroforestry. For example, low soil fertility could be addressed by various technologies such as fertilizer, livestock manure, green manure from trees or shrubs, crop rotations. Next, each alternative is evaluated for its technical potential and suitability to farmers resources and capabilities, infrastructure, and support services. A judgement is then made on what interventions seem to have the highest potential. Agroforestry interventions are proposed only when they have a comparative advantage.

If agroforestry seems viable, the list of high-priority constraints will suggest specific options for consideration. For example, a fodder shortage problem could be addressed through the following seven agroforestry options.

a) Establishment of fodder banks for a cut and carry system

b) Improvement of grazing management through live fencing

c) Introduction of fodder trees for browse in grazing lands

d) Planting of fodder trees with grass in intensive feed gardens

e) Planting of MPTs / grass strips along contours in crop land

f) Establishment of alley cropping mainly for fodder production.

"Ex-ante" evaluation of a technology is a part of technology assessment. It is carried out to determine a technology's potential for adoption. Ex-ante evaluation means the evaluation of the likely impact of proposed technology before the technology has been introduced. It is based on appropriate assumptions using relevant data from other sources. This requires knowledge of technology management and performance under the specific conditions of the LUS. If a proposed technology is well known and some farmers are familiar with its management and requirement then a recommendation for extension programme can be formulated. On the other hand, if very little is known about the technology, then the D&D team needs to propose research activities. A programme of research is proposed to develop specific components, understand technical relationships, and/ or to test/ adapt the technology or components.

f) Definition of agroforestry research needs

A comparative analysis of the research needs for each agroforestry technology for each LUS within an eco-zone is to be carried out. This analysis will be the basis for the design of appropriate research programme. Thus the main output of the macro D&D exercise is the definition of a research agenda to develop relevant technologies for the ecozone of interest.

g) Inter-institutional participation/ coordination

The results of macro D&D will suggest specific problem areas for complementary research in different institutions and better use of their scientific and physical resources. Multi-institutional participation in strategic phases of the research, such as D&D exercise, definitely facilitates the integration of individual effort and the development of coordinated programme.

13.2.2 Micro D&D

13.2.2.1 Objectives of micro D&D

The objectives of micro D&D are similar to those of macro D&D. The major difference is that macro D&D has a broad scope (i.e., an ecozone) whereas micro D&D has specific zone. In micro D&D, since resources are inevitably limited, a country, institution or project will have to be selective in deciding which land use system should be subjected to micro D&D. The choice of the land use system for micro D&D depends on the following criteria

a) Political and economic importance of the system

b) Technical potentials for improvement of the LUS and

c) Scientific expertise and other resources in the natinal collaborating institutions for carrying out research in the LUS.

13.2.2.2 Analysis of land use system and constraints

This phase of micro D&D aims at

a) Prioritizing the needs of the household

b) Identifying production constraints (both those that can and cannot be manipulated) and

c) Assessing potentials for system development.

The basic framework used for this analysis is a farming system, where the decision making unit is the household.

The household usually manages a combination of crop, livestock, and tree production systems, along with other non-agricultural and off-farm activities, to satisfy its basic felt needs of food, cash, fuelwood, building materials, and security. Besides endogenous factors, the farming system is influenced by exogenous factors of a political, social, economic, or technological nature. Understanding the interactions within the farming system and the effects of environment is essential for prioritizing the needs of the household, identifying production constraints, and assessing potentials for system development.

The micro D&D research team, therefore, has to quantify resources, management, and yield of each component of the farming system, including characteristics and priorities of the household. The constraints analysis aspect of micro D&D is well suited for planning the initial stages of research. However, later stages of research may require a reassessment or a more precise measurement of some of these constraints.

13.2.2.3 *Design & evaluation of agroforestry technologies*

The design and evaluation objective of micro D&D focuses mainly on the following.

a) Technology specification and
b) Ex-ante evaluation of technology

In technology specification, for any type of production system, whether crop, livestock or agroforestry, a technology can be defined as a "package" of husbandry practices and inputs which is specified in terms of i)the farming systems/household it is targeted to; i) its components and resource requirements; iii) the management and implementation regimes to be followed by the farmers; and iv) the estimation of real benefits and costs to the farmers under favourable and unfavourable conditions.

In other words, technology specification should provide sufficient detail to permit technical feasibility analysis, socioeconomic analysis and assessment by farmers. Because agroforestry is a new science, there is a dearth of technical information on most components. For examples, information on the biophysical productivity of MPTs under different arrangements and management regimes is known only for a few species in selected environments. Similarly, not much is known about utilization and timing of MPTs outputs for crop and livestock productivity in alley cropping. Successful technology design requires adequate research experience.

Ex-ante evaluation of technology is simply the analysis of its probably impacts and implications. The analysis looks at benefits and conflicts or problems likely to arise at the levels of farming system, community or-village and region or catchment area.

The ex-ante analysis should assess the production potential and technical feasibility of the technology. There are four essential types of analysis involved in ex-ante evaluation, viz.,

a) *Economic viability:* Benefit/ cost ratio, net returns to land / labour / cash, risk and sensitivity analysis.

b) *Sustainability:* Analysis of the technology's capacity to meet objectives in short and long terms; also analysis of expected changes and requirements related to soils, water, vegetation, management and commercial input / output streams.

c) *Farmer acceptability:* It is essential to analyse who in the household makes decisions on the resources required, who has to do the work, and who will receive the benefit accruing from proposed changes.

d) *Adoption potential:* Analysis of technology impacts in terms of number of farmers, regional development priorities, tenure rights, institutional and infrastructural support systems, etc.

During the design process, the scientists should interact with typical households of the target system and with extension and development agents, particularly in relation to the following topics viz., i) Priority problems being addressed and expected performance levels of the technology, ii)Endogenous factors and constraints to successful adoption iii) Resource and management requirements for effective establishment (transitional analysis) and iv) Expected benefits, and impacts on farmers objectives.

13.3 Design and Evaluation of Research programme

There are four types of scientific research, viz.,

a) Basic research which is designed to generate new knowledge or understanding.
b) Strategic research to solve specific research problems.
c) Applied research to create new technology and
d) Adaptive research to adjust technology to the specific needs of a particular set of biophysical or socio-economic conditions.

In the case of agroforestry, where basic and applied research is not well developed and where farmers have more experience than scientists with management of technologies. On-farm research may have stronger role to play in the research strategy, including applied research.

13.3.1 Research Design Criteria

Agroforestry technology must be specified at least in its principal components viz., MPT species, spatial arrangement, management regimes, and performance levels. Different types of trials are conducted to achieve these specifications. Within the ICRAF D&D scheme, the three general categories of agroforestry trials are i) General MPTs screening trials, ii) MPTs technology screening trials and iii) MPTs management trials. It is expected that any D&D exercise will provide sufficient understanding of the farmers environment and production systems for designing the varied research programme.

13.4 Methodological considerations in D&D

a) Research team

The nature of agroforestry systems suggests the need to set up multi-disciplinary and multi-institutional teams to achieve the objectives of D&D. For macro D&D, efforts should be made to form teams comprising.

a) Biophysical scientists from the fields of soils, crops, climate, livestock and forestry and

b) Social scientists from the fields of agricultural economics, rural sociology, or anthropology

For micro D&D and follow up studies, the expertise and composition of the team are based on the prioritized LUS, its identified constraints and the specific objectives of the study. Team leadership is critical for successful D&D implementation. An objective of D&D strategy is the key to bring disciplines to work together to address the problems of the farmer in an integrated manner.

b) Research domains and recommendation domains

Defining the target land use system or farming system is probably the most crucial step in the agroforestry systems research process. The problem is that unless researchers have developed the technology and knows how to manage it, what requirements it has and exactly what it can do, they do not have a solid basis to define an appropriate recommendation domain. Accordingly, the research team should always be concerned about whether and how technology development is modifying the original research domain.

Precise definition of the target system is also important for the reason that every farming system or household is somewhat different. Some customizing or fine-tuning of technology is required for each case; this is what occurs in the adoption process. Thus the definition of research domains should strike a wise balance not too general so as to be useless as a guide to research, nor too specific so as to apply only to a small number of farms or households.

The research team can assess the effect of its work on the research domain by answering the following questions.

a) Which LUSs are most affected by the problem under investigation?

b) Which systems can benefit (one hopes the majority) and to what extent can they benefit from the technologies being developed?

c) What are the major conditioning and determinant factors, endogenous or exogenous to the farming system, for technology management and performance?

d) What are the expected benefits and impacts of technology adoption?

c) Data collection methods

The D&D methodology employs several data collection methods appropriate to its specific objectives. Each method has its own strengths, weaknesses, degree of reliability of collected data and resource requirements. D&D exercises should generate a minimum of raw data, a maximum of useful information, and in a timely manner. For this reason, the preliminary D&D work will take a rapid appraisal approach using secondary information surveys. D&D work at later stages in the research and development process will use methods that make a maximum contribution within the limits of available human and physical resources.

d) Analytical methods and the role of farmers

The major decisions in the D&D analysis are derived using interactive and holistic methods with the principal actors being the D&D team and the farmers / households. This interaction should be based on solid information, consultation with development agents and policy-makers, and a commitment to arrive at logical conclusions in the process.

To ensure effective participation in discussions, all participants must show mutual respect and accept that each can make a valuable contribution. This should be reflected in observance of the following guidelines of behaviour.

a) Understand a point from the other's perspective.
b) Criticize constructively, admit if wrong, stress the positive.
c) Reason – don't argue.
d) Explain thoroughly.
e) Offer helpful suggestions and
f) Avoid snap judgments.

However, this does not mean that participants should accept everything said. There is a need to challenge, seek clarification, and discover the root causes of disagreements or conflicts. In this respect, the farmers should be treated as equal participants.

e) Logistical and operational aspect

For macro D&D, the total staff time required from planning to conclusion is estimated to be roughly three months. An approximate breakdown of the work schedule could be as follows.

i) Planning the study and orientation of the entire team from 2 to 3 days.

ii) Review and synthesis of secondary information from 2 to 3 weeks.

iii) Field work from 2 to 3 weeks depending on the geographical size of the ecozone.

iv) Final analysis and reporting from 1 to 1.5 months.

For micro D&D, the total input of staff time and logistic support is approximately similar. However, a formal survey (i.e. a survey of 5-100 farmers with a semi-structured questionnaire) will normally constitute an important part of the study.

This framework is claimed to be applicable to both research and extension activities. If the agroforestry technologies that are envisaged in the design already exist, the D&D methodology can be used directly as a guide for agroforestry interventions by the extension workers. If, on the other hand, the desired technologies do not exists or are not fully developed, the design can provide a basis for identifying the kind of research that need to be undertaken. This will prove to be a most applicable to the developmental project with direct involvement of the researcher and the farmer at one platform.

13.5 Criteria of Good Agroforestry Design

A good agroforestry design should fulfill the following criteria

a) *Productivity:* There are many different ways to improve productivity with agroforestry viz., increased output of tree products, improved yields of associated crops, reduction of cropping system inputs, increased labour efficiency, diversification of production, satisfaction of basic needs and other measures of economic efficiency or achievement of biological potential.

b) *Sustainability:* By seeking improvements in the sustainability of production systems, agroforestry can achieve its conservation goals while appealing directly to the motivation of low income farmers, who may not always be interested in conservation for its own sake.

c) *Adoptability:* No matter how technically elegant or environmentally sound an agroforestry design may be, nothing practical is achieved unless it is adopted by its intended users. This means that the technology has to fit the social as well as environmental characteristics of the land use system for which it is designed.

Questions

1. A family of procedures for the diagnosis of land management problems and potentials and the design of agroforestry solution is

   a) Agroforestry D&D b) Sustainability
   c) Productivity d) None of the above

2. One of the important key features of agroforestry D & D is

   a) Meeting objectives b) Problem solving
   c) Flexibility d) All the above

3. Agroforestry D & D covers entire ecological zone within a country is

   a) Mini D & D b) Micro D & D
   c) Macro D & D d) None of the above

4. Agroforestry D & D covers within the chosen ecological zone is

   a) Mini D & D b) Micro D & D

   c) Macro D & D d) All the above

5. The main criteria for good Agroforestry design are
   a) Suitability and compatibility
   b) Compatibility and productivity
   c) Productivity and sustainability
   d) None of the above

References

Atul. 2004. Agroforestry - Diagnosis and Design (D&D) methodology. In: Methodologies in agroforestry research. Eds. Rai, P. and A.K. Handa. All India Coordinated Research Project on Agroforestry, NRCAF, Jhansi. p. 1-14.

Hegde, N.G. and V.D. Kelkar. 1989. Agroforestry – selected readings. BAIF Publication, Pune. 107 p.

Huxley, P.A. and P.J. Wood. Undated. Technology and research considerations in ICRAF's "Diagnosis and design" procedures. ICRAF Working Paper, 26, Nairobi.

Jha, L.K. 1995. Advances in agroforestry. APH Publishing Corporation, New Delhi. 669 p.

Pathak, P.S., H.M. Pateria and K.R. Solanki. 2000. Agroforestry system in India. Diagnosis and Design approach. ICAR, NRCAF, Jhansi. 228 p.

Raintree, J.B. 1987. D&D user's manual. Nairobi, ICRAF.

—•—

CHAPTER **14**

Economic Analysis of Agroforestry Systems

Agroforestry must be more profitable than monocropping land use system. There should be biologically and economically advantageous interactions among its individual components. In economic terms, agroforestry land use systems should have a higher output value at the same resource cost and / or the same output value at a lower resource cost. Hence economic analysis of the agroforestry systems should be done to help the farmers to make decisions about the allocation of scarce resources in a rational way in order to meet the designated objectives.

14.1 Economic Analysis at different times

The economic analysis at different times can be grouped into two types as below.

i) Ex ante economic analysis

ii) Ex post economic analysis

14.1.1 Ex Ante Economic analysis

The main aim of ex ante economic analysis of a potential agroforestry system is to analyze and escalate prior knowledge and idea of its technical feasibility and economical viability of the system. It may also help to prioritize the various technology options available and/ or guide scientists to formulate or propose other technologies.

14.1.2 Ex Post Economic analysis

The main aim of the ex post economic analysis of an agroforestry is to determine whether the system should be readjusted to the specified objectives. This analysis is served as yard stick for measuring the system whether it is progressing on its way or not. The flaws can be rectified and possible to improve the efficiency of the system.

14.2 Economic Analysis from different view points

The economic analysis at different view points can be classified into two types as below.

i) Private economic analysis
ii) Public economic analysis

14.2.1 Private Economic analysis

Private economic analysis is referred as financial analysis. In private economic analysis, the objectives of the individual or group are to maximize the income with minimum cost and risk. This is reflected in the inputs and outputs to be considered as well as the values given to them. The cash flow and consumption pattern of the individual can be known by this analysis.

14.2.2 Public Economic analysis

Public economic analysis is referred as economic analysis and that it takes into consideration the costs and benefits outside the farm or area where the system is present. This analysis considers the true economic value (shadow price) of inputs and outputs of the system. This method analyse the welfare of the whole / entire society in that particular area.

14.3. *Evaluation Criteria*

The agroforestry systems can be evaluated by using the following tools.

a) Net Present Value (NPV)
b) Benefit Cost Ratio (BCR)
c) Internal Rate of Return (IRR)

a) Net Present Value

The net present value of AF system is the present value of all expected cash flows which is obtained by present worth of cost stream from the present worth of benefit stream. The NPV should be positive for any system

$$NPV = \sum_{t=0}^{n} \frac{(B_t - C_t)}{(1+r)^n}$$

where,

Bt - AF system benefit stream in period 't'

Ct - AF system cost stream in period 't'

r - Rate of interest or discounted rate

n - Number of years for which the system will operate

Where 'B' is the benefits in a year 't'

'C' is the costs in a year 't'

b) Benefit Cost Ratio

BCR is defined as the present value of benefit stream is divided by the present value of cost stream.

$$BCR = \sum_{t=0}^{n} \frac{\dfrac{Bt}{(1+r)^t}}{\dfrac{Ct}{(1+r)^t}}$$

Where 'B' are the benefits accruing in year 't'

'C' are the costs accruing in year 't'

'r' is the selected discount rate

'n' is number of years for which the system will operate

The BCR should be greater than one for agroforestry system, then only it can be considered worthy.

c) Internal Rate of Return

IRR calculates the maximum rate of interest that a project can repay on loans while still recovering all investment and operating costs. The IRR determines the earning power of the money invested in a particular venture.

$$IRR = \text{Lower discount rate} + \text{Difference between the higher and lower discount rate} \times \frac{\text{NPV at lower discount rate}}{\text{Absolute difference between NPV at two discount rates}}$$

14.4 Types of Economic analysis

a) *Labour input analysis:* it shows the flow of labour inputs required for the introduction and maintenance of an agroforestry technology.

b) *Material input analysis:* it shows the flow of material inputs required for the introduction and maintenance of the AF system

c) *Cash flow analysis:* It shows the flow of cash expenditures and receipts results from the introduction and maintenance of AF in the system. Such analysis should include loans and repayments when applicable.

d) *Discounted cost / benefit analysis:* It determines the profitability of the agroforestry system.

14.5 Data requirements for Analysis of Agroforestry systems

i) Inputs: Quantitative Aspects

Inputs are called as *resources or production factors* or goods and services used to produce an output. The main inputs are land, labour and capital which include buildings, equipment, livestock, seeds and fertilizers.

a) *Land* should be specified in hectares, tenure status, productive capacity and how it is used over time.

b) *Labour input* is expressed in work days or hours by type, activity, and timing during and over the years.

c) *Capital inputs* are specified in weight, volume, or numbers by type (planting material, fertilizers, agricultural chemicals, water, tools, fodder, equipments, etc. and other sources (farm produced, purchased or hired).

ii) Outputs: Quantitative Aspects

Outputs are goods or services produced by an activity. Major goods are crops and livestock products, timber and fuel wood. Service outputs are live fence, soil conservation works and nitrogen fixation, etc., The outputs need to be specified in terms of units such as volume, weight, numbers, quality and timing.

14.6 Valuation of Inputs

In financial analysis, inputs are valued at their market price or at their opportunity cost to the individual concerned. In principle, market prices are applied when inputs are commonly purchased or hired by the individual. The opportunity costs are used where most of the inputs are provided by the farm family.

In public analysis, the shadow price of the input, representing the true economic worth to society as a whole, has to be used.

a) Land

When a farmer wants to change the existing land use to an agroforestry land use, there is no need to value such land. If a farmer has to purchase or rent a land to introduce the agroforestry system, such purchase or rental price has to be entered into the financial analysis.

b) Labour

In financial analysis, all labour hired is valued at its market price while all family labour is valued at its opportunity cost.

c) Capital

In financial analysis, all purchased materials and equipments are valued at their market price.

14.7 Valuation of Outputs

Similar to the inputs valuation, outputs are valued at the market price or opportunity cost in private economic analysis and at shadow price in public economic analysis.

a) ***Agricultural products:*** Outputs are valued at the market price or opportunity cost in the private economic analysis and at shadow price in the public economic analysis.

b) ***Fuel wood:*** In the private economic analysis, fuel wood is valued at the local market price if commonly sold and/or purchased by the farmers concerned. To value the standing volume of fuel wood, the price to be used should be the net of the cost of labour for collecting it and transport cost. If neither is the case, fuel wood is valued either in terms of alternative fuels used by the farmers or in terms of labour savings made. A frequently used alternative fuel source is kerosene and sometimes dried cattle dung.

c) ***Tree Fodder:*** Tree leaves and pods have increasingly been recognized as potential sources of animal fodder, especially in arid and semi-arid conditions. In private economic analysis, tree leaves may be valued at market price if they can be sold locally. However, if leaves are not sold, they can be evaluated on the basic energy or protein value. Deriving the shadow price of tree fodder for a public economic analysis requires adjustment of market and opportunity prices, subsidies and taxes.

d) ***Leaf Litter:*** Leaf litter from trees and shrubs may be used to add nutrients and organic matter to the soil. So far, there are no recorded instances of leaf litter being sold commercially. The market price may be derived, however, on the basis of nutrient content and prices of commercially available fertilizers.

14.8 Environmental Outputs

i) Preservation of productive capacity of a resource.

ii) Improvement of the productive capacity of a resource.

iii) Preservation /Improvement of the productive capacity of a resource.

Examples of AF technologies with environmental output are tree based soil and water conservation measures, soil improvement through nitrogen fixing trees, wind protection through shelterbelts and preservation of forest areas or woodlands.

In private economic analysis only direct (environmental) benefits considered whereas in public economics analysis, all benefits (both direct and indirect) are considered.

The valuation of environmental outputs may be handled in different ways as below.

i) Difference in output stream of the production system affected by the environmental activity

ii) Cost of preventing / reducing the damage done to the environment due to the lack of an environmental activity

iii) Purchase price of the land affected by the environmental activity.

Questions

1. Which analysis gives an idea of whether a system is technically feasible and economically viable

   a) Ex post economic analysis b) Ex ante economic analysis

   c) Both of the above d) None of the above

2. Which analysis indicate whether a system should be readjusted so as to better meet the objectives

   a) Ex post economic analysis b) Ex ante economic analysis

   c) Both of the above d) None of the above

3. Private economic analysis is otherwise known as

   a) Financial analysis b) Economic analysis

   c) Public analysis d) None of the above

4. Public economic analysis is also known as

   a) Financial analysis b) Economic analysis

   c) Public analysis d) None of the above

5. The tools of agroforestry system evaluation are

   a) B:C ratio b) Net present value

   c) Internal rate of return d) All the above

6. Which tool determines the earning power of the money invested in a particular venture

   a) B:C ratio b) NPV

   c) IRR d) None of the above

7. Market prices are applied when

   a) Inputs are commonly purchased or hired by government

   b) Inputs are commonly purchased or hired by individual

   c) Inputs are commonly purchased or hired by private

   d) Inputs are commonly purchased or hired by institution

8. In case of tree fodder, valuation may be given to it as

   a) Oil content b) Nutrient content

   c) Energy / protein content d) None of the above

10. The analysis in which loans and repayments should be included to show the flow of cash expenditure and receipts.

    a) Discounted cost / benefit analysis b) Material analysis

    c) Labour analysis d) Cash flow analysis

References

Anon. 1994. Economic modeling for agroforestry in Tamil Nadu. Centre for Research in Economic and Social Development. 196 p.

Chundawat, B.S. and S.K. Gautham.1996. A text book of Agroforestry. Oxford and IBH Publishing Co. Pvt. Ltd., New Delhi. 188 p.

Dwivedi, A.P. Agroforestry – Principle and Practices. Oxford and IBH Publishing Co. Pvt. Ltd., New Delhi. 365 p.

Howard A. Steppler and P.K.R. Nair. 1987. Agroforestry – a decade of developments. ICRAF, Nairobi. 335 p.

William R. Burch and T. Kathy Parker. 1992. Winrock International, U.S.A. and South Asia Book, U.S.A. 183 p.

—•—

CHAPTER 15

Silviculture of Important Agroforestry Trees

15.1 *Teak*

Scientific Name	:	*Tectona grandis*
English Name	:	Teak
Tamil Name	:	Thekku
Hindi Name	:	Sagwan
Family	:	Verbenaceae

15.1.1 Distribution

Native to Southeast Asia (India, Myanmar, Thailand and Western Laos). Teak (*Tectona grandis*), is the most important of the three species in the genus (*T. hamiltoniana, T. phillipensis*). A rare combination of durability, dimensional stability and strength properties make teak a paragon of timber.Its latitudinal limits are 9°N - 25° N and its longitudinal limits are 70°E - 100° E. In its natural habitat, teak occurs in mixed deciduous forests normally constituting a small number of individuals but occasionally in pure stands. Since 1840 it is raised as plantations in India and Myanmar.

15.1.2 Physiognomy

It is a large, deciduous tree, attains a height of 30 to 40 m under favouable conditions.On good sites, clean bole of 15 to 30 m length are obtained. Fluting and buttresses are often found at the base of the trees. Bark is thick, grey or lightish brown, fibrous with shallow longitudinal fissures, peeling off in long thin narrow flakes on older trees. Leaves are large (25 - 50 cm in length and 15-35 cm in width), elliptic or obovate; upper side green to dark green in colour; underside has dense whitish to tawny mat of wooly hairs. Leaf arrangement is opposite. The flowers are small, whitish and appear in large panicles containing few thousand flower buds which open only a few at a time during the flowering period of 2 to 4 weeks. The fruit is a hard, irregularly rounded drupe containing 4 seed chambers. The pericarp consists of a thin papery exocarp, a thick felty, brown mesocarp and a stony endocarp. Only rarely do all the four seed chambers contain developed seed. Generally in a sample, one seeded fruits account for 42-64%; 2 seeded account for 12-25%; 3 seeded vary from 2 to 6% and 4 seeded accounts for 1-2%. Large numbers (11-35%) are also found to be completely seedless. The teak fruit varies from 11 to 18 mm in cross section and an average of 2000 seeds are present in a kg (Plate 9).

15.1.3 Phenology

Age of trees at first flowering varies markedly depending on site, climate, silvicultural management and genetic linkage. In its natural habitat, it comes to flowering in 6-8 years. But under plantation conditions, size or height of the trees rather than age exercises profound influences on flower initiation. Flowering generally occurs during June - July. The fruits attain full size and mature in about four months after fertilization (i.e. during October- November). A sign of maturity is their facile fall to the ground when the tree is agitated.

Plate 9. *Tectona grandis*

15.1.4 Silvicultural Characters

Teak seedlings are sensitive to frost and drought. It is a strong light-demander, intolerant of suppression by weeds. It is a fire resistant; seedlings and saplings killed back by fire and frost. It coppices and pollards vigorously, upto about middle age.

15.1.5 Climate and Soil

The tree grows under a wide range of climatic situations from sea level upto an altitude of 1200m and in a rainfall range of less than 900 mm to more than 2500 mm.The most suitable soil for teak is the deep, well drained alluvium having relatively high contents of calcium and phosphorus. It tolerates a pH range of 6.5 to 8.0 but good growth is attained on soils of pH 6.5 and an annual rainfall of 1500 mm. Teak is a pronounced light demander and does not tolerate suppression at any period of its growth. It is also fairly fire tolerant.

15.1.6 Nursery Technique

The seeds that have been stored for at least one year germinate better than fresh seed. The seeds are subjected to alternate wetting and drying of 24 hr duration each for seven days.Use of large seeds more than 14mm in diameter gives better germination. Seeds are sown @ 1 kg m^{-2} on raised nursery bed of 10m long, 1 m wide and 0.5m in height and covered with soil to a depth equivalent to the fruit diameter.

To prevent soil erosion, beds are reinforced on the sides with bamboo splits or other such material. Germination commences in about 15 days, accelerates during the next 15 days, declines thereafter. Majority of the seeds germinate in 40 days. The germination percent is around 40 per cent. The beds are watered twice daily for the first two months, once daily for the next three months and on alternate days thereafter. Super sized seedlings called "wolf' smother others. For field planting, only stumps prepared from one year old seedlings are used as it grows faster. Stumps are prepared by cutting away from the seedling everything except 2.5 cm of the shoot and 22.5 cm of the root. The stem portion receives an oblique cut and the root portion a horizontal cut. All laterals from the tap root are pruned away. The stumps should preferably be planted within 2 to 3 days.

15.1.7 Planting

Stumps are planted at a spacing of 2 m x 2 m in crow bar pits. The initial plant density of 2500 ha^{-1} is reduced in a phased manner to 80 to 100 ha^{-1} by an operation called thinning. A total of four thinning is given in 5th, 10th, 18th and 28th years and at each thinning, an existing population is reduced by half. The first two thinning are mechanical and are done according to a rule of thumb. In the first thinning, alternate diagonal rows are removed; in the second thinning alternate rows are felled. The 3rd and 4th are silvicultural thinning in that they are restricted only to diseased and malformed trees. Trees possessing clean, cylindrical and straight bole coupled with less taper, small crown and less fluting are retained. Final felling is done at the end of 60 years. A single tree yields 1.5 m^3 of timber. Its rotation is 40-60 years.Its yield is about 6500 cu.ft of stem wood per acre.

15.1.8 Utilization

Teak wood is globally renowned for its strength, durability, dimensional stability, working quality and non-corrosive property when in contact with metal. The durability is attributable to the deposition of polyphenols in its heartwood. On account of these outstanding properties, Teak is called as the Queen of timbers. Increasingly large quantities of teak are used by the plywood industry for high grade commercial and tea-chest categories of plywood. Lops and tops and other rejects serve as fuel wood. Teak leaves are often used as platters.

15.2 *Eucalyptus*

Scientific Name : *Eucalyptus tereticornis*
English Name : Eucalyptus hybrid
Tamil Name : Thailam
Family : Myrtaceae

15.2.1 Distribution

It is native to Australia.*E. tereticornis* was first introduced in the Nandi hills (Karnataka) by Tiuppu Sultan between 1782 and1790. Now it is grown over one lakh ha in Peninsular India. Extensive plantations have been raised to meet the needs of fuel wood, small timber and pulpwood in Punjab and Haryana, where area under forest is negligible.Large scale plantations of the species were taken up in Uttar Pradesh from 1962 onwards. In Tamil Nadu, it is being planted extensively for pulp and paper.

15.2.2 Physiognomy

It is a tall tree with stout trunk, attains a height of 50 m. It is an evergreen, glabrous tree usually secreting an aromatic gum. The leaves and flowers contain conspicuous oil glands. Leaves of the saplings are generally opposite, sessile, cordate and held horizontal; those of the adult tree as a rule are alternate, petiolate and held vertical. Flowers are borne in umbels usually pedunculate. Calyx tube encloses the ovary which is covered with a deciduous operculum.The operculum is much longer than calyx and is formed by the union of the petals and falls off entire when the stamens emerge. Flowers are white in colour. Fruit consisting of the enlarged calyx tube is usually hard and woody, full of resin sacs. Seeds are numerous but a large proportion of the seeds is abortive and sterile seeds outnumber fertile ones. Bark is grey, exfoliating in long flakes (Plate 10).

15.2.3 Phenology

It flowers almost throughout the year.The capsules are collected six months after anthesis when they just turn dark brown. If the capsule is left for long on the tree it will burst and shatter the seeds. Hence capsules are collected and kept in trays / tarpaulins. After sun drying for 3 or 4 days, the empty capsules are removed. Mature seeds are dark brown / black in colour. The seeds retain viability for up to 5 years.

15.2.4 Silvicultural Characters

It is a light demander. It produces good coppice freely and vigorously. It is a fast growing species and adaptable to a wide range of soil and climatic condition.

15.2.5 Climate and Soil

It grows up to an altitude of 500 m. It is sensitive to frost. It grows in alluvial, black cotton, gravelly, lateritic, skeletal rocky and murram soils and even on shifting sand dunes. In highly calcareous, very saline and alkaline soils, clay and kankar pan the growth is limited. Deep, fertile and well drained loamy soil gives best growth.It is suited for the plains receiving a rainfall of 800-1000 mm. The tree prefers sandy loam to loamy soils within a pH range of 6.0 - 7.5.

15.2.6 Nursery Technique

Seeds are sown in raised beds of 1 x 1 x 0.15 m. After wetting the bed, sieved seeds @ 5 g m^{-2} are mixed with a small quantity of sand and evenly spread on the bed. The seeds are covered with a layer of soil. The nursery beds need to be kept moist by watering twice daily. A mulch of hay prevents leaching of seeds during watering. Watering

Plate 10. *Eucalyptus tereticornis*

is done through a fine rose can. BHC 10% dust has to be applied on the bed to prevent ants and termites. Five and 10 days after sowing, 2% copper fungicide must be applied. Germination starts on the 5th day. Since seedlings are sensitive to extreme sunlight, the bed has to be protected by a shade screen during the first fortnight. Thirty days old seedlings are gently lifted from the bed and containerized in 200 gauge polybags measuring 20 cm x 10 cm. The polybags are filled with 2:1:1 mixture of red soil, sand and FYM. After wetting the filled polybags, seedlings are pricked out and transplanted in polybags. The transplanted seedlings should be provided shade for a week and watered twice a day. Six month old seedlings are used for planting. The containers must be shifted once in every fortnight from the second month to prevent rooting.

15.2.7 Planting

The seedlings are planted at a spacing of 2 m x 2 m in pits of 30 cm^3. Quality of seedlings is determined by the thickness of the root collar region than by height. The trees are felled at the end of five to seven years depending on the utility.Thereafter two coppices are taken at intervals of five year each. Coppice management is important in eucalyptus. Hundreds of new shoots develop on the margin of the cut stem. Felling of the trees prior to or immediately after the monsoon helps in rapid callus formation and thicker coppice shoots. Care should be taken to fell the trees with a gentle slope at the cut so that rain water does not collect as a pool and cause decay of the callus tissue. Though hundreds of coppice shoots develop yet only four to five stems ultimately remain on the stump and the others are edged out in natural competition. However in man made plantation, the coppice shoots are thinned leaving two dominant shoots in order to increase the productivity of tree. The productivity of coppice plantation is generally higher by 20-25 % than the first seedling rotation. At the end of the second coppice,it is necessary to uproot the tree. The productivity of rainfed plantations in Tamil Nadu plains ranges from 50 -75 ton per ha at the end of seven years.

15.2.8 Utilization

Eucalyptus wood is the main raw material for paper industry in Tamil Nadu. Currently, it is used for making packing cases and 70% of the requirement in Himachal Pradesh for apple transport is met by this species. Leaves contain oil. Bark yields oxalic acid. It is preferred by the farmers by virtue of several characters like (i) fast growth; (ii) non browsable; (iii) immunity to pests and diseases; (iv) good coppicing ability.

15.3 Tamarind

Scientific Name	:	*Tamarindus indica*
Tamil Name	:	Puli
Hindi name	:	Imli
Family	:	Caesalpinae

15.3.1 Distribution

Tamarind is native to dry savanna of tropical Africa. In ancient times, it was introduced to Asia by Arab traders. In Tamil Nadu, it is being extensively cultivated in Dharmapuri, Morappur, Krishnagiri, Anjatti and Hosur areas.

15.3.2 Physiognomy

Tamarind is a medium to large sized evergreen tree, up to 24 m in height and 7 m in girth. The bark is brownish or dark grey and fissured longitudinally and horizontally. Leaves are paripinnate and 15 cm long. Leaflets vary from 10-20 pairs; oblong. Flowers are borne in small terminal, drooping racemes on current season's growth.

Pods are 7.5 - 20 cm long, 2.5 cm broad and 1 cm thick, more or less constricted between the seeds, slightly curved, brownish - ash in colour. A tree in Urigam Village (Dharmapuri) possesses long pods measuring 30-45 cm. There are 3 to 12 seeds in each pod which are obovate, oblong, compressed with a shallow, oblong pit on each side of the flat face, smooth dark brown and shining. Seeds are contained in loculi, enveloped by a tough, leathery membrane called endocarp. Outside the endocarp is the light brownish, red sweetish, acidic, edible pulp, traversed by a number of branched, ligneous strands. The pod is shell, fragile and easily separable (Plate 11).

15.3.3 Phenology

Tamarind flowers from April - July in most areas of South India. Peak flowering occurs during May - June. New leaves appear in May and are closely followed by the flowers. Fruits mature during winter season in South India.

15.3.4 Silvicultural Characters

It is a light demander; very sensitive to frost; drought resistant. It is deep rooted and wind-firm; slow-growing. It coppices fairly well; produces root suckers freely.

15.3.5 Climate and Soil

It is a tree of tropical climate, tolerating temperatures up to 47^{0}C but is very sensitive to frost and fire. It prefers mean annual rainfall of 500 to 1500 mm and tolerates water logging, but also grows well with 350 mm annual rainfall. It can be grown under a variety of soils ranging from gravelly to deep alluvial. It thrives best on deep alluvial soil with adequate supply of water.

15.3.6 Nursery Technique

Pods are collected during February - April. The fruits are dried in sun; the outer shell is removed by hand or by

Plate 11. *Tamarindus indica*

beating with a mallet; the seed is separated from the pulp by hand and washed in water. Washed seeds are dried under shade and stored. Seeds from crown collection are superior to those of ground collection.About 1800-2000 seeds weigh one kg.

Seeds are normally sown directly in sand medium or polybags. Germination takes place in 5-10 days and is completed in 30 days. Young plants grow fast on porous soil and a soil mixture of 1: 1: 1 red soil, sand, farm yard manure.

A low cost technology of soaking for 24 hr in a solution prepared by dissolving handful of cow dung or cow's urine results in vigorous seedling to the extent of 75- 80 per cent.

15.3.7 Planting

One year old seedlings are planted at a spacing of 10m x 10m. A full grown tree yields 180-225 kg of fruit per tree.

15.3.8 Utilization

All plant parts find some use, but the most useful is the fruit which contains sweetish acidic pulp. The Tamarind of commerce which is widely used for souring curries, sauces, chutneys and certain beverages. The pulp is also employed in medicine. Leaves boiled along with gingelly oil is applied to relieve swelling caused by sprains / fractures and also to relieve pain. Pulp mixed with sugar and made into tamarind balls is used for seasoning other food. Refreshing acid drink and syrup are also made. Leaves serve as good fodder. Tender leaves, flowers and young seedlings are used as vegetable. Seeds are eaten after roasting or boiling. Powdered seeds are used as cattle feed. Processed seed powder is used in confectionery. Kernels contain a polysaccharide having very good sizing properties and extensively employed as a source of sizing powder in cotton and jute industries. The tree yields timber, hard and difficult to work. The tree is extensively used for avenue planting.

15.4 *Ailanthus*

Scientific Name	: *Ailanthus excelsa*
English name	: Tree of Heaven
Tamil name	: Pimaram, Pinari, and Perumaram
Hindi name	: Arna, Ardu, Maharukh, Arusa
Family	: Simarubiaceae

15.4.1 Distribution

Ailanthus excelsa is considered a native of the Indian peninsula, but occurs throughout the tropical and subtropical regions of India, especially in the dry districts of Gujarat, Rajasthan, Haryana, Punjab, Uttar Pradesh, Bihar, Orissa and the Deccan plateau. It is not found in the high rainfall regions of the West Coast.

15.4.2 Physiognomy

It is a large deciduous tree, attains a height of 18-24 m. Bark is light grey and smooth in young trees, with large conspicuous leaf scars, rough, granular and greyish brown in older ones. Leaves are pinnately compound, up to 1m long, with 8-14 pairs of leaflets and an unpleasant smell when crushed. In seedlings and saplings leaves are imparipinnate for the first three or four years, after which the terminal leaflet becomes reduced in size or is represented by a mere prolongation of the rachis and finally the typical abruptly paripinnate leaves are formed about the fourth or fifth season. The fruit is a red one-seeded samara, 5-7.5 cm long and 1.2-1.5 cm wide, prominently veined, acute at both ends and twisted at the base; about 9 to 10 fruits weigh one g. The fruit being winged is adapted for dissemination by wind (Plate 12).

15.4.3 Phenology

The panicles of small yellowish flowers appear in February - March and the fruits ripen in May - June. Old leaves fall during February and new ones appear in March - April. In Central India, flowers appear during February and March and in the North during April.

15.4.4 Silvicultural Characters

It is a strong light demander and is susceptible to frost and prolonged drought, though poles and trees are resistant. It coppices well and produces root suckers freely. It is very susceptible to water logging and wind break.

15.4.5 Climate and Soil

It grows well in semi-arid and semi moist regions, both in the plains and in the hills. In Rajasthan, it grows in areas with an average annual rainfall of 400 mm. It avoids moist areas with high monsoon rainfall. The average minimum temperature is about 10°C and maximum about 30°C. Average annual temperature is about 27°C. Though it can grow on a wide variety of soils, it thrives best on porous sandy loam. It avoids clay with poor drainage and waterlogged areas. It can be grown on shallow dry soils but growth is poor.

15.4.6 Nursery Technique

For maximum viability and vigour, fruits are collected when the colour of the pericarp changes from yellowish brown to brown. The samaras are dried in the sun. They lose their viability in about four months. Nicked samaras are kept in wet gunnies for 48 hour. At the end of storage, sprouted samaras are separated and sown in polybags. By this method, the number of empty containers is considerably minimized.Samaras number about 9600 in a kg. Germination takes about 5-12 days and is completed in about 30 days. The radical and plumule emerge through the winged covering, the cotyledons being carried up and the testa usually left inside the fruit. The hypocotyl arches somewhat at first, soon straightening. The cotyledons usually persist for about 2-3 months after which they turn yellow and fall.

Plate 12. *Ailanthus excelsa*

15.4.7 Planting

Nine month old seedlings are planted at a spacing of 5m x 5m. Seedling growth is fairly fast attaining a height of 0.2, 0.6, 2.4 and 4.2 m at the end of first, second, third and fourth growing seasons respectively.

15.4.8 Utilization

A ten year old tree approximately yields 50-75 tonnes per hectare of match wood. The wood is soft, white, very light but fairly strong and easy to saw. It is used for match splints, packing cases, fishing catamarans and floats. It is also used for commercial plywood. The wood is perishable in the open but not under water. It is grown as shade and avenue tree in hotter parts of India. It yields an inferior type of gum. Its bark and gum are of medicinal value.

15.5 Neem

Scientific Name	: *Azadirachta indica*
English name	: Neem, Nim, Indian Lilac, Chinaberry
Tamil name	: Vembu, Veppamaram, Veppan
Hindi name	: Bal-nimb, Neem, Neim, Nimb, Nind
Family	: Meliaceae

15.5.1 Distribution

The tree is believed to be a native of upper Myanmar. Its occurrence in the Siwalik forest of Uttar Pradesh is also considered natural. In India, neem occurs in tropical dry deciduous and thorn forests in drier parts up to 1500 m. It is found in almost all states of India.

15.5.2 Physiognomy

A large evergreen tree attains a height of 12 to 15 m and occasionally upto 25 m with a clear bole of 3-7.5 m,

1.8-2.8 m girth. It branches early and forms a broad rounded crown of bright green foliage. Bark is moderately thick with scattered, small tubercles between numerous longitudinal and oblique wrinkled furrows, dark grey outside and reddish inside. Leaves are imparipinnate, 20-38 cm long, crowded near the end of branches, leaflets 9-13 nearly opposite, oblique, lanceolate, some times falcate, acuminate, deeply and sharply serrate, glabrous on both surfaces, petioles very short. Inflorescence is axillary, many flowered panicle, shorter than the leaves. Flowers are white, fragrant, shortly pedicelled and bisexual. Fruit is a drupe, 12-18 mm long, ovoid-oblong, smooth, dark yellow or purple when ripe, endocarp thin cartilaginous, intercellular spaces appear between the epicarp and the endocarp; their wall break down and form the mucilaginous pulpy mesocarp. Seeds 1-2, reticulate (Plate 13).

15.5.3 Phenology

Neem is almost evergreen but becomes near leafless in dry localities for a short period during February-March. New leaves appear in March-April, before the old ones are shed. Flowering occurs from January-March; in the southern parts of India. In Kerala flowering starts in January, in Karnataka, Tamil Nadu and Andhra Pradesh during February-March; in Central India during the first week of April. Thus flowering is progressively delayed from South to North in the sub-Himalayan area where flowering occurs during the first week of May. Fruiting also follows the pattern of flowering; fruits ripens from June to August. The tree starts fruiting at the age of five years but economic yield of fruits is obtained at the age of 10-12 years. About 3300-4500 seeds weigh one kg and on an average, a medium sized tree produces 37-55 kg fruits.

15.5.4 Silvicultural Characters

It is a light-demander. It is sensitive to frost and fire. It is drought-hardy.

Plate 13. *Azadirachta indica*

15.5.5 Climate and Soil

It grows on almost all kinds of soils including sandy, clayey, saline, alkaline, black cotton soils and laterite crusts. However, it does not grow on salty flats, inundated areas and soils with finely divided mica.The tree has a very wide range of climatic adaptability and grows well in areas with a mean annual temperature ranging between 0°C and 45°C. It is a light demander but survives under shade. The rainfall in the areas of its occurrence varies from 450 to 1150 mm.

15.5.6 Nursery Technique

Physiologically mature seeds with maximum germination capacity and longevity are obtained 10-12 weeks after flowering. Ripen fruits are collected by beating the branches and heaped in the shade by mixing ash, till the seed becomes easily extractable. Seeds are dried in shade for 3-4 days and then sown or stored. Neem seeds do not require any special pretreatment but seeds do not retain viability very long and have to be sown within 2 or 3 weeks after collection. Depulping and cleaning of seed improve the germination per cent and seeds passed through digestive system of birds show better germination. Seeds are sown in the nursery beds at a depth of 2.5 cm .Germination normally takes one or two weeks and it may vary from 9 to 55 days. The beds should be sparingly watered and soil kept loose and excessive water should be avoided.

Seedlings are pricked out to transplanting beds at 15 x 15 cm spacing or to polypags when they are about 5 cm high; neem seedlings do not require shade except during pricking out stage. In frosty localities, the plants should be provided with shelters. Seeds can also be sown in polybags directly.

15.5.7 Planting

Polybag seedlings or root-shoot cuttings are more successful for agroforestry, silvi-pastoral and roadside avenue plantations. One year old seedlings are preferably planted at a spacing of 5m x 5 m.

15.5.8 Utilization

For centuries the neem tree, has provided man with twigs for tooth brushes, pharmaceuticals for aches and pains and pest control agents against insects. The drought tolerant neem helps to reduce soil erosion and produces soap, lamp oil, lubricant and lumber. It is also a good shade tree. Every part of neem has its own importance.

Neem wood is moderately heavy, stable and resembles mahagony in appearance; is resistant to fungi and most borers. It is used for making furniture, cart axles; yokes, naves and felloes, boards and panels, cabinets, packing cases, ornamental cuttings, ship and boat building, helms, oars, oil-mills, cigar boxes, carved images, toys, drums and agricultural implements. The leaves are palatable to cattle and buffaloes and constitute a traditional feed in several parts of the country. In Andhra Pradesh, these are regularly fed to cattle and goats to increase secretion of milk.They are carminative and aid digestion; leaves also used as mulch and manure. Tender leaves in combination with *Piper nigrum* are found to be effective in intestinal helminthiasis. Neem oil and its derivatives are mainly used in soap making and in preparation of tooth paste. Oil is a remedy in some chronic skin diseases and ulcers and is externally applied for rheumatism, leprosy and sprain. Warm-oil relieves ear trouble, cures dental and gum troubles and provides relief in asthma when taken with betel leaf. Oil is reported to have anti fertility properties and it possesses antiseptic and antifungal activity.

15.6 *Pungam*

Scientific Name : *Pongamia pinnata*

Tamil Name : Ponga, Pungam

Hindi Name : Karanj, Kanji

Family : Leguminosae

15.6.1 Distribution

Pongamia pinnata grows throughout the greater part of the country chiefly along streams and rivers and in beach and tidal forests. It is considered to be a native of the Western ghats, ascending upto an elevation of about 1200 m. It is a characteristic tree of mixed forests in the Andaman. As a scattered tree it grows in Sub-Himalayan tract also, It has been widely planted in different parts of the country.

15.6.2 Physiognomy

Full grown *Pongamia pinnata* is a moderate sized tree with a short bole and spreading thick crown. Branches are stout. Leaves are imparipinnate, glabrous and light green leaflets are opposite, without stipules, 5-12 cm long and shortly acuminate. The bark is thin, smooth, grey and yellowish inside. Pods are indehiscent, turgid almost woody pointed at both ends, 4-5 cm long, and seeded. Seed is compressed, wrinkled, reddish brown and oily (Plate 14).

15.6.3 Phenology

Pongamia pinnata is almost an evergreen tree exception in dry locatitites where it becomes leafless for a short period. Flowering takes place during June.

15.6.4 Silvicultural Characters

It is a drought resistant; can withstand frost. It is shade tolerant but does well with full over head light. It is an excellent coppicer; puts forth root suckers readily. It is frequently pollarded in South India for green manure.

15.6.5 Climate and Soil

Its wide distribution in the country shows that *P. pinnata* can adapt itself to a spectrum of climatic conditions.

Plate 14. *Pongamia pinnata*

In areas where it has been successfully grown, the absolute maximum shade temperature varies from about 27 to 38°C and rainfall from about 500-2500 mm. It can grow on a wide variety of soils, ranging from sandy to black cotton soils, but it grows best on deep and well drained alluvial soils with abundant supply of soil moisture.

15.6.6 Nursery Technique

Pod collection is done by beating the branches or from ground. The pods are dried in the sun and thrashed to extract the seed, which is dried in shade before storage. Seed yield per tree varies from 9 to 90 kg. Soaking of seeds in cold water for 24 hr hastens and improves germination. About 1200 seeds weigh one kg. Seed length varies from 1.3 cm to 2.30 cm and seed breath from 0.90 cm to 1.3 cm. Seeds are sown in drills 15 cm apart and at a depth of about 2 cm.

Germination starts in about 10 days and takes about a month to complete. The percentage of fertility is high and a germination percentage of about 60 to 80 can be expected. One kg seed will produce about 1000 plants. *Pongamia pinnata* can be raised by direct sowing or by planting out of entire plants or stumps. Branch cuttings can also be used.

15.6.7 Planting

One year old seedlings which attain a height of about 50-60 cm are planted out. Entire plants or stumps may be used for planting; the use of the latter is more convenient for large scale plantation. For making stumps, plants of about 1-2 cm collar diameter are preferred. Spacing adopted is 5m x 5m. For avenue planting a spacing of about 8m is adopted. Pits of 30 cm^3 are dug for planting.

15.6.8 Utilization

Leaves of the trees are lopped for fodder, even though their digestibility is poor. Green pods are also fed to cattle in Maharashtra. The seed cake after oil extraction is used in poultry ration. The oil is used in soap manufacture and as a biofuel. The wood is a good fuel.

15.7 *Prosopis*

Scientific name	:	*Prosopis juliflora*
English Name	:	Mesquite, Mesquite bean
Tamil Name	:	Velikaruvel, Semaikaruvel, Delhimul
Hindi Name	:	Vilayati babool, Vilati khejra
Family	:	Mimosae

15.7.1 Distribution

Mesquite tree is native to Central America and South America. The tree has been planted in many arid zones of the world. It is widely propagated in Africa and Asia, particularly in India. *P.juliflora* was first introduced into India from Mexico in 1877 from seeds obtained through Kew Botanical Gardens, England. It occurs throughout the districts of Tirunelveli, Ramanthapuram, Virudhunagar in Tamil Nadu. Sparse occurrence of P. *juliflora* is restricted to stream courses and road margins in Pudukottai and Thanjavur districts. *Prosopis juliflora* has proved to be a great social asset at no cost to the Government. It has brought back vast stretches of land into use and labour into productive employment in Tamil Nadu.

15.7.2 Physiognomy

Five forms of *Prosopis juliflora* are recognized in India viz., Mexican, Australian, Argentinean, Peruvian and the Arid; the first two are more common. Leaves are alternate on the branchlets of current year and fascicled on those of the previous year; 3 to 13 cm long with petioles enlarged and often glandular at the base; rachis 2 to 13 cm, long slender terete or slightly winged, prolonged beyond the last pinnate as a soft bristle leaflets 8-48 pairs, 5 to 50 mm in length and 4 to 5 mm in breadth, linear or oblong reticulately veined. Stipules linear, deciduous, branches are long, zig-zag struggling or pendulous, sometimes armed with scattered stout, subulate thorns, 1 to 5 cm long and often marked with minute dark lenticels and irregularly shaped red blotches. Bark is thick,dark, reddish brown, divided by shallow fissures. Flowers are creamy white, in axillary pedunculate spikes, 4 to 10 cm long, solitary or in fascicles of 2-4; peduncles 5 to 18mm long. Pods are linear, compressed when young subterete at maturity, straight or falcate, contracted at the two ends usually with a long stout, base; pale yellow or straw coloured when ripe; mesocarp pulpy; endocarp cartilaginous, surrounding each seed separately. Seeds number 12 - 34 in a pod, oblong, flattened, with thin light brown lustrous testa and thin thorny albumen (Plate 15).

15.7.3 Phenology

The Australian variety of *Prosopis juliflora* flowers and fruits in India twice a year. First flowering commences during September- October and may continue up to the end of February. The second flowering which is profuse occurs during February - March. Pods ripen in November - January and again during April -June. The interval between flowering and fruiting is nearly 1 to 1½ months. Trees begin to bear fruit in the third year and sufficient quantity of seed is available from the fifth year. Almost every year is a good seed year.

15.7.4 Silvicultural Characters

It is a strong light-demander, hardy, drought-resistant, fast growing. It is a frost -tender. It coppices well and produce root-suckers freely.

Plate 15. *Prosopis juliflora*

15.7.5 Climate and Soil

It is the only exotic species capable of growing on a wide variety of soils including heavy clays, black soil, red loam, coastal sandy soil, alluvial soils, saline and neutral soils, inland as well as loamy soils. It comes up in low as well as high rainfall areas, in hills and plains alike. A hot, dry climate with mild winter is preferred by the species. In areas where it has been naturalized, the absolute maximum shade temperature varies from 40.6° C to 45.6°C. It comes well under a rainfall regime of 150 to 750 m; it can be grown even in areas of still lower rainfall of 70 mm per year. It does not tolerate heavy rainfall.

15.7.6 Nursery Technique

Ripe pods are collected from trees by lopping branches or from the ground, and sun - dried. The fruits turn yellow when mature and fall to the ground. When dry, the pods can be broken into segments by beating with a mallet. Pounding also breaks the pods into bits, each bit containing a seed. It is very difficult to extract the seed from the dry broken bits as the endocarp is hard and does not yield to mechanical pressure. Protracted soaking in water, drying and beating can separate the seeds and then it can be stored for about two years. As the germinative capacity of the seed is diminished due to attack by some insects, the pods are better fumigated with carbon disulphide or hydrocynic acid before storage. The seeds treated with naphthalene powder and stored in sealed tins may retain 60% viability at the end of 5 years. Seeds stored in gunny bags retain only 36% viability after 3 years. Naphthalene is harmless to man and treatment with this chemical is safer. Storage in bins with 5 cm layer of sand on top is stated to prevent entry of beetles. The number of seeds per pod varies from 4 to 32. Seed number varies from 32,000 to 35,000 in a kg. The germinative capacity of clean treated fresh seed is 80-90%. The seed is very hard with an impermeable seed coat and requires pretreatment to hasten germination. The seed requires any of the following treatments (i) soaking the seed in cold water for 48 hr. (ii) soaking the seed in commercial H_2SO_4 for about 20 min; iii) soaking the seed with boiling water, allowing it to cool for 24 hr. After pretreatment, the seeds should be sown in polythene containers filled with soil mixture.

15.7.7 Planting

When the seedlings attain a height of 20-30 cm, they are ready for planting. As the tree has a tendency to become branchy, pruning is of importance when it is grown in avenues and all shoots except one may be cut down. Leading shoots occasionally require mechanical support for the first one or two year. In 3-4 years, the leading shoot makes headway and takes the pleasant appearance initially of a well grown shrub and eventually of a tree. The trees are cut once in three years. The first felling promotes good coppice growth. The second felling is at the end of sixth year. If the growth is very good and if immediate cash is desired the interval may be reduced to two years. The rootstock is generally uprooted at the third felling. At every cutting an approximate yield of 10-15 ton ha^{-1} can be expected.

15.7.8 Utilization

It is a versatile multipurpose plant; used as fuel, fodder, and several utility timber and therefore described as "Loyal Timber of the poor" *Prosopis juliflora* wood is moderately hard and moderately heavy (specific gravity 0.83 to0.88 air dry and 0.78 oven dry). It is excellent for firewood and makes superior charcoal. Because of its high calorific value, the wood has been termed "Wooden anthracite". It burns slowly and evenly and holds heat well. The wood is very durable and is used for fence posts, door and window frames and other light carpentry. The wood is used in making pulp and paper. It is mainly used in brick kiln. Prosopis is the main fuel for pottery, cottage industries also. Heart wood is durable and is suitable for tool handles of all sorts, crushers and such other purposes where both strength and hardness are of importance. The flowers are a source of high quality honey. The mesquite pods are sweet, succulent and edible, having a high nutritive value. The pods are eaten by livestock (sheep and goats) and also made into flour for human consumption during famine.

Small young leaflets are also eaten by cattle, sheep and goats. Foliar analysis revealed that the magnesium and phosphorus contents are less than 1.0% and sodium is high indicating the plants tolerance to salinity. The bark is a source of tannin; it also produces gum and is used for sizing cloth. The roots yield a fibre for cordage and coarse fabric. It also yields a gum which is used for adhesive purposes. Its wood is used for production of writing and printing paper. However, the yield is low and the strength property of the pulp is poor as compared to most other hardwoods.

15.8 Casuarina

Scientific Name	: *Casuarina equisetifolia*
English Name	: Beaf wood
Tamil Name	: Savukku
Hindi Name	: Janlisaru
Family	: Casuarinaceae

15.8.1 Distribution

Casuarina is an exotic to India. It was introduced in Karwar District in 1668. It occurs naturally in Andaman, Bangladesh and Burma coast. It is planted extensively in coastal areas of Tamil Nadu and other coastal states.

15.8.2 Physiognomy

Casuarina is a large fast growing evergreen tree with graceful appearance, resembles a feathery conifer. Bole is long and cylindrical. In its natural state it is gregarious, forming pure crops with little or no under growth except grass and sporadic shrubs. The tree attains height upto 40m and a diameter of 60cm, often buttressed at the base. It is short lived; its natural span of life seldom exceeds 50 years. In less favourable localities, it turns misshapen and hollow beyond 25 years of age (Plate 16).

15.8.3 Phenology

The tree flowers generally twice a year, during February- April and September-October. Fruits (cones) appear in June and December. Variations in flowering and fruiting may occur with localities.

15.8.4 Silvicultural Characters

Casuarina is a fast growing, light demanding species. It is very sensitive to excess soil moisture, fire and frost. It does not tolerate water logging for long. It does not tolerate drought up to sapling stage, later with deep rooting, it can withstand drought. It tolerates low temperature and shade. As a general rule it does not coppice but can withstand pollarding. Rare instances of natural regeneration and root suckers are noticed. It improves soil fertility by virtue of its vigorous root nodulation with nitrifying bacteria.

15.8.5 Climate and Soil

Along the coastal regions, where Casuarina tolerates the extreme temperature up to 47°C. In peninsular India, the rainfall is between 900 to 3800 mm. Growth is poor in low rainfall areas.

Casuarina grows best in loose, fine coastal sands. It can grow well in inland in well drained sandy soils. It

Plate 16. *Casuarina equisetifolia*

tolerates lateritic and red soils and also saline, alkaline and acidic conditions. The trees have nitrogen fixing root nodules and hence dependability on nitrogen supplement is less. Casuarina is essentially a tree adapted to coastal plains and has been cultivated in inland terrain. The altitudinal range can be taken from sea level to 1300 m.

15.8.6 Nursery Technique

The nursery should be located as close to the planting site as possible. The area should be plain, with sandy or sandy clay soils, with reliable water source. The area should be cleared off all the debris and miscellaneous growth if any. A light ploughing should be carried out before alignment. The beds should be aligned in north-south direction. Beds of 10 m ×1 m is generally followed. The aligned beds are dug up to depth of 30 cm. The cavity so formed is filled with debris or leaf litter and burnt.

In deep sandy soils, the dug up earth is mixed with Farm Yard Manure at 1:1 ratio and the cavity is filled and made compact, the beds are formed either to ground level or sunken to 5 cm depth.

In sandy clay soils, the beds are formed with the dug up soil mixed with sand, FYM in equal proportions filling the cavity. The bed is formed above the ground level raised to a height of 20 to 25 cm. While forming the beds, parathion dust at ¾ kg per bed or any ant repellent should be mixed with the soil to prevent ants carrying away the seed. Root nodules or soil from older plantations can be mixed with the soil in the beds.

Each bed of 10 sq.m is sown with 400 to 500 gm. of clean seed. The seed should be mixed with fine sand and sown broadcast on the beds. They should be covered lightly with hay and watered twice a day. Watering should not lead to water logging; otherwise damping off will result. In case damping off is noticed a light spray with blitox dissolved in water is to be done.

In about ten to fifteen days, the germination will be completed. Light watering twice a day is to be done and there after daily once for a month or till the seedlings reach 10 cm height. In case of casualties due to termite attack, parathion dust or chlorpyriphos should be applied.

Secondary beds are made to transplant the seedlings germinated in primary beds. The area is ploughed lightly before forming the beds. A bed of 10 m × 1 m is dug up to a depth of 30 cm. Dry leaf debris is spread in the cavity and burnt to sterilize the same. The cavity is then filled with soil mixture of local sand and FYM at 1:1 ratio. The beds are formed to a height of 9" to 12" in deep clayey soils. In sandy soils, the beds are formed to ground level. The beds are mixed with any ant repellent and blitox.

A steel frame of 1 m width thick nails of 3" length welded at 5 cm spacing, there will be 22 nails in each row and 4 such rows are made in each frame. After watering the beds, the frame is pressed to give exact alignment for the seedlings to be planted on the beds. The seedlings after treatment should be planted in the holes and a suitable shade material or with palm leaves is provided for a period of one week to ten days.

The beds should be watered daily, till the seedlings attain 45cm, which will be possible in about 45 days to 60 days.Planting in secondary beds must start in April and completed, so that the seedlings are of optimum height in June-July. Planting in beds in north east monsoon conditions must be done in July.The beds should be weeded regularly at least till the seedlings are beyond damage. With the onset of rains in July/October, the seedlings should be taken out for planting.

The calendar of operations differs depending on the rainfall pattern. The nursery life is four to six months, starting from sowing in primary beds. If the planting is in July, the work should start in January-February, if it is in September-October the work should start in A1pril May.

The planting stocks are of the following types.

i) *Naked seedlings obtained from primary beds* sown simply by broadcast of seed. Most of the farmers use such seedlings when they are over one meter high. The survival and growth is very poor. Some success is seen when planting is done during good rains.

ii) *Naked seedlings obtained from secondary beds.* The seed is sown in beds primarily to get juvenile seedlings up to 10 cm height. They are pricked out and planted at a spacing of 5 cm × 5 m on secondary beds, which when grow to height of 35 to 50 cm, in about two months, are planted out. The survival is better and with good rains, they establish and grow well.

iii) *Bag Culture:* The seedlings from the primary beds are pricked out and transplanted in polythene bags filled with soil mixture and watered. These seedlings have good success and establish quickly and grow well. Soil mixture of sand and FYM at 1:1 ratio is well mixed, filled in polybags and compacted. The bags are arranged in form of beds at 1000 in each and watered thoroughly.

Seedlings of equal height from primary beds are taken out; the long thin roots are cut with a sharp knife. The roots are then dipped in a sticky mud paste, with addition of gum to increase the stickiness, so that the muddy paste will cover the root system. These seedlings are planted in the center of the bags and watered. If the beds are in open land, shade with coconut leaves at 8 to 10 feet. The shade is provided for a period of one week to ten days, to avoid casualties. The shade is removed when the seedlings are established. Shifting of bags is done when roots start protruding, while doing so; the seedlings are to be graded according to height. When the seedlings attain a height of 45 to 50 cm, they are ready for planting.

iv) *Treated naked seedlings:* Seedlings raised in secondary beds are taken out and the thin roots are cut cleanly. The roots are then treated by immersion in a sticky mud puddle, to ensure stickiness of the soil paste, a small quantity of gum is added. These when planted during rains, are as good as bag plants. They are very cheaper in cost and easy to handle.

15.8.7 Planting

Three to four month old nursery raised seedlings have to be planted. The plantation site in sandy soils is cleared of all the debris, roots and stumps if any, are removed. A light ploughing with country plough is carried out. In clayey sandy soil the ploughing should be deep with a heavy plough. Regarding spacing, normally farmers adopt 0.8 to 1 m spacing. 2 m × 1 m spacing allows inter cropping in the first year and this also gives good growing space for the trees to put on good girth at an early stage.

Cluster planting, by planting two seedlings at each point is also practiced.

For planting of bag plants in sandy soils, the soil is scooped to the depth and width of the bag, then planted during onset of rains. If watering is done, planting establishment will be faster. In hard clayey soils, a small pit of 30 cm is dug out and bag planting is done after removing the polythene cover carefully, without disturbing the soil in the bag.

Bag culture involves careful handling of the bags, during loading, transport and unloading, as the soil mixture in the bag is loose. Faulty handling will expose the seedlings and the benefit of raising bagged seedlings will be lost.

Planting of naked seedlings or treated naked seedlings can be done in holes made with a crowbar. The seedling is placed with roots straight in the hole. Before planting the thin part of the root should be cut cleanly with a knife.

15.8.8 Utilization

a) Pulpwood

Casuarina is a good pulpwood species. Farmers are cultivating mostly this species for pulpwood. Yield and strength properties of Casuarina pulp are reported satisfactory for wrapping paper and duplex paper. It makes good pulp by use of neutral sulfite semi chemical process. The stem with bark is used for pulping.

b) Fuel wood

Casuarina is considered to be the best firewood in the world, burns even when green. Its calorific value is 4950 cal/kg.

c) Timber

The sapwood is pale brown, heartwood dark reddish brown. Timber is strong and heavy (Average 850 kgs/ m^3). It is liable to crack and split, not easy to saw and season. It is used as poles, scaffolding, transmission lines and rafters.

d) Medicinal value

Bark of Casuarina is a tonic and astringent useful in dysentery and diarrhea. Decoction of leaves and twigs is used in colic and powdered seed is made into a paste and applied as balm for headaches.

e) Tannin and Resin

Bark contains 6-18% tannin, which is also used for dyeing wool and silk fabrics and for toughening fishermen nets. It also yields a resin.

e) Avenue

Casuarina forms a good avenue tree and most suitable for landscaping sea beaches. It forms a good hedge plant and can be shaped to desired form.

f) Windbreak

Casuarina with deep taproot can withstand cyclonic storms than any other species and is very useful as a windbreak and for sand dune stabilization.

g) Soil improvement

Casuarina develops nitrogen fixing bacterial nodules of Frankia species on the roots. They fix considerable quantities of nitrogen in the soil and help to improve the soil nutrient status. Casuarina in sterile sandy soils at a age of 12 years, over 27 metric tonnes of leaf litter and humus from needles get added to the soil. The insects, micro fauna and flora help in improving the soil conditions.

15.9 Silk Cotton

Scientific Name : *Ceiba pentandra*
English : Silk cotton
Tamil : Elavam
Hindi : Semal
Family : Bombacaceae

15.9.1 Distribution

It is very widely distributed, almost throughout India, excepting the arid region; ascending up to 1200 m, occasionally even up to 1500 m. It is typical of alluvial savanna type of forests; it grows sporadically in mixed deciduous forests of West Coast, and the evergreen forests of Bengal and Assam. It is widely distributed in peninsular India. Indo-genetic plain and parts of the sub-Himalayan tract.

15.9.2 Physiognomy

It is a huge tree with a clean bole. Its wood is very soft and light. Untreated wood is highly perishable. It coppices in early years and produces root suckers (Plate 17).

Plate 17. *Ceiba pentandra*

15.9.3 Phenology

Leaf-fall occurs in the month of December. The leaves turn yellow and fall off. The flowering takes place in the leafless period from January to early March. The fruits occur in April and May. Seeds are surrounded by floss and are hence dispersed by wind.

15.9.4 Silvicultural Characters

It is a strong light demander and drought resistant. Fairly resistant to frost and fire. It produces root suckers and early coppicer.

15.9.5 Climate and Soil

Absolute maximum and minimum temperature is 35°C-49°C and -4° to 18°C respectively. Normal annual rainfall is 750-2500 mm. Best suited soil is alluvial soil having a considerable proportion of sand and good moisture supply.

15.9.6 Nursery Technique

Seeds are sown in the nursery bed in lines 15 cm apart. When it reaches two leaf stage, it is pricked out and transplanted in to polybags. Regular watering and weeding are to be done. Six months old seedlings are ready for transplanting.

15.9.7 Planting

In kapok, nursery raised seedlings or stumps prepared out of 1-2 years old plants are planted at 5 m x 5 m to 8 m x 8 m spacing.

15.9.8 Utilization

Its main use is flosses. It is used for stuffing mattresses and pillows. Its wood is light and soft and hence recommended for matchwood.

15.10 Black Babul

Scientific Name : *Acacia nilotica*

English Name : Indian gum arabic

Tamil Name : Karuvelam

Hindi Name : Babul

Family : Mimosae

15.10.1 Distribution

Babul is indigenous to the Western part of the Indo-Gangetic plains and the Northern part of Deccan plateau, including Andhra Pradesh, Maharashtra, Rajasthan and Gujarat. It is widely planted, or self sown throughout the hot regions of India, viz., Punjab, Haryana, Uttar pradesh, Madhya pradesh and Karnataka. It is an important constituent of southern dry mixed deciduous tropical forests, northern and southern tropical thorn forests of India at an elevation of 200-500 m.

15.10.2 Physiognomy

It is a moderate - sized, almost evergreen tree with a short trunk, a spreading crown, and feathery foliage. Bark is dark brown, nearly black, pinkish brown and hard inside, with regular deep longitudinal fissures which very often run spirally up the tree. Young branches green, pubescent. Stipulate spines straight, white, up to 2 inches long, variable, sometimes absent in old trees. The tree varies much in size, attaining a height of 50 to 60 ft. or even more, and occasionally a girth of 8 to 10 ft (Plate 18).

15.10.3 Phenology

The young leaves appear from March to May, the old leaves commencing to fall before they appear. Flowering is most general in the rainy season, from June to September,

Plate 18. *Acacia nilotica*

but trees may be found in flowers as late as December or January. The time of fruit ripening varies according to locality, but is usually from April to June.

15.10.4 Silvicultural Characters

It is drought-resistant; frost-tender. It is strong light-demander. It's coppicing power is very variable, generally poor.

15.10.5 Climate and Soil

A nilotica can grow on a variety of soils, provided sufficient moisture is available. It prefers well drained fresh alluvial sandy loam soil in riverain tracts, though it can grow on clay and black cotton soil also. It can stand mild soil salinity provided sufficient moisture is available. In its natural habitat, average rainfall is 400 to 1500 mm; fairly drought resistant, but thrives best in areas with 500-1250 mm.

15.10.6 Nursery Technique

Babul seedlings are raised in polythene bags. Treated seeds are sown, about 1.5 cm deep, 2-3 seeds in each bag in February - March and regularly watered and weeded. Excessive watering should be avoided; shading is necessary to avoid surface cracking.

15.10.7 Planting

For obtaining bigger plants, seeds is sown in June - July in bigger bags and one year old seedlings are planted out. It's rotation is 30 years for timber and 15-20 years for tannin; it yields 23.02 m^3 wood per ha. at the age of 30 years and 8-10 tonnes of pods per ha.

15.10.8 Utilization

Leaves and pods are widely used as fodder. It is an extremely valuable source of fuel wood and charcoal of excellent quality. General utility of timbers for construction of carts, wheels, agricultural tools and implements, doors, windows, mine props, fencing materials, etc. Bark is one of the best tanning materials. Babul gum is used in inks, paints, matches and confectionery.

15.11. Australian Wattle

Scientific Name : *Acacia auriculiformis*

English Name : Australian Wattle

Tamil Name : Kathivel

Family : Mimosae

15.11.1 Distribution

It is native of New Guinea, North Australia and Queensland and is exotic to India. It has been successfully raised in West Bengal, Bihar and Andhra Pradesh and also in Karnataka, Orissa, Uttar Pradesh, Maharashtra, Kerala and Tamil Nadu.

15.11.2 Physiognomy

It is a moderate sized tree, attaining a height of 10m sometimes up to 15m. Bark is smooth, white or light grey, fissures occur in later years. The leaf stalks are modified into flattened blade called as phyllode, which is narrow, oblong, slightly curved and sickle shaped, leathery and dull green (Plate 19).

15.11.3 Phenology

Seeds ripen during January - March; best time for collection of pods is February - March, by lopping and spreading out for 5-9 days in the open area. Seeds retain viability up to 2 years. Germination capacity is about 50%.

Plate 19. *Acacia auriculiformis*

15.11.4 Silvicultural Characters

It is a strong light demander but regenerate under shade. It is a fire tender. It is drought tolerant but sensitive to severe drought. It produces profuse root suckers but is a poor coppicer under Indian condition.

15.11.5 Climate and Soil

It grows from sea-level to about 700 m altitude. It does well in temperature of 26-30°C and can survive up to 40°C. It is well adapted to drought and grow with as little as 600 mm rainfall.

15.11.6 Nursery Technique

Seed requires pretreatment, either 24-48 hours immersion in water at room temperature or in cooling boiled water, for 24 hours. It may be sown in nursery beds or in polythene bag containers during March to early April, under shade which is removed after germination is completed. Seed is sprinkled with kerosene oil to protect against ants. Germination starts in 5-10 days and complete in 20-25 days.

15.11.7 Planting

Four months old nursery raised seedlings are planted out with ball of earth, when 15-30 cm tall, at the break of monsoon in June - July. It can be successfully raised by stump planting also. Planting is done in pits of 30 cm^3 to 45 cm^3at a spacing of 1 m x 1m to 2m x 2m.

15.11.8 Utilization

It is reasonably good fuel wood, furniture wood and pulpwood. It is suitable as a shade and ornamental avenue tree and also for afforestation and reclaiming wastelands.

15.12 Bamboos

15.12.1 Introduction

Bamboo is a vernacular term for members of subfamily bambusoideae of family Poaceae. Almost all the 75 genera and 1250 species of bamboo are woody and fast growing species. Bamboo is a cultural feature of south-east Asia. No country in the region is without indigenous bamboo flora. Its plethora of essential uses has led to the use of terms such as "Bamboo Culture", "Green Gold"," Poor man's timber" Friend of the People" and "the Cradle of Coffin Timber".

15.12.2 Distribution of Bamboos

Bamboos are versatile group of plants of multiple utility and usually form rich belt of vegetation in moist deciduous, semi-evergreen, tropical, subtropical and temperate forests. Asia is rich in bamboos with approximately 65 genera are endemic to the region. The number of species is around 900 which spread about 14 million hectares of earth surface and 80 per cent of bamboo distribution occurs in Asia.

India shelters a large number of bamboo species and is considered as one of the largest reserves of bamboos in the world. There are 125 indigenous and exotic species of bamboos falling under 23 genera in an area of 10.03 million hectares. They form rich belts of vegetation in well drained parts of tropical and subtropical habitats and rise up to 3700 m altitude in Himalayas. More than 50 per cent of bamboos occur in eastern India viz. Arunachal Pradesh, Assam, Manipur, Tripura, Meghalaya and West Bengal. The principal bamboo genera occurring in India are *Arundinaria, Bambusa, Dendrocalamus, Indocalamus, Semiarundinaria* and others.

15.12.3 Description

a) Rhizome

The basal portion of the culm, containing very short internodes, which grows almost horizontally and remains at level to the soil surface or little below it is known as rhizome. According to the pattern of growth basically,there are two types of rhizomesviz.,i) sympodial, also called as pachymorph, and the other monopodial, also called as leptomorph.

Sympodial rhizomes are solid, usually short and thick, and their lateral buds only produce new rhizomes, each of which eventually produces a solitary culm. The sympodial arrangement ensures that each new culm will have its own new roots to supply its needs. The constricted proximal portion of the rhizome which develops first is known as "neck". The short necked rhizome system develops dense clumps. Comparatively longer necked rhizome system develops loose clumps in which culms are not tightly packed. The longer the rhizome neck, the higher the capacity for the vegetative spread. For instance, *Melocanna baccifera* which successfully invades large areas in open habitats often forming pure stands has rhizome neck areas as long as one metre. The genus viz., *Bambusa, Dendrocalamus, Gigantochloa, Schizostachyum* and *Thyrsostachys* have sympodial rhizomes.

b) Culm

The buds on rhizome nodes enlarge for many months in the soil and tender shoots (sprouts) emerge as pointed cones completely covered with imbricate sheaths from 50 to 60 days. Rain in May - July strongly induce sprouting. These shoots elongate rapidly and in about 3-4 months the young culm grows to its full height before branching. In giant bamboo, *Dendrocalamus giganteus,* the culms may reach a height of 35 m and diameter of 30 cm whereas in the smallest bamboo *Arundianaria pygmaea* the culms do not exceed half a metre height and 1 cm diameter.

Most of the bamboos have hollow culms: very few have solid culms eg species of *Dinochloa.* In the case of bamboos having hollow culms the walls may be thin or thick. The culms may be erect or arching outward in most of the bamboos.

c) Culm sheath

These are modified leaves arranged in two ranks mainly for the protection of the growing culms. The sheath is attached at the node, covers the culm, and when it becomes old falls of leaving a scar. Width of the sheath is greater than the circumference of the nodes so that the two sides of the sheath overlap. The back of the sheath is usually partly or wholly covered with hairs which are easily detached when sheaths are old. The inner side of the sheath is smooth and shining. The sheaths are arranged alternately in two opposite ranks.

15.12.4. Flowering

Flowering is either periodic or annual. It is either gregarious or sporadic. Gregarious flowering is usually followed by the death of clumps. The gregarious flowering proceeds from one end of the forest to another. Most bamboos flower at long intervals.

15.12.5 Natural Regeneration

Natural regeneration of bamboos occurs after a gregarious flowering. Under natural conditions, the seed germinates at the commencement of the rainy season. The natural seedlings in flowered bamboo areas take the place of the flowered crop in about 8 -10 years. The young seedlings do not tolerate heavy shade and are liable to be killed by a heavy growth of weeds, but light shade is helpful in protecting them from drought and frost.

15.12.6 Artificial Regeneration

Seed is propagated though (i) Direct sowing, (ii) Raising seedlings in the nurseries and transplanting. In direct sowing, seeds are sown directly. For raising seedlings, seeds are raised in nursery beds, then pricked out seedlings are planted in the field. Bamboos are also propagated by (i) rhizomes, (ii) cuttings and (iii) layering.

The culms of two or three rhizomes attached are planted. The stem cuttings and rhizome cuttings are planted. For layering, the culm is partly cut through near the base and bent down below the ground, which would eventually result in development of roots and shoots at the nodes. The internodes are then cut through and the resulting plants are separated and planted out.

15.12.7. Management of Bamboo Plantations

Management of bamboo plantations is easier than management of bamboo in natural forests. As management of bamboo forests did not receive adequate attention in the past, the important factors for consideration in management of bamboo forest are (i) Felling cycle (ii) Intensity of felling and (iii) Method of felling.

The general tendency is to fix 3 year felling cycle. While fixing the felling cycle the age at which the culm becomes exploitable and the age at which it dies are also taken into consideration. Intensity of bamboo felling also varies from place to place.

The following felling rules are adopted for bamboos.

i) Restriction on cutting of one year and sometimes even two years old culms
ii) Restriction of some older bamboos for support of immature culms.
iii) Prohibition on digging of rhizomes.
iv) Regulation of the height at which bamboo should be cut. ie.15 cm.

v) Insistence on cutting with a sharp instrument so that the stump does not split.

vi) In case of flowering, the bamboo should be cut only when the seed has been shed.

vii) Cleaning and tending of clumps have to be done to facilitate growth of new culms.

Among the various species available in the country, the following two bamboos viz., *Bambusa bambos* and *Dendrocalamus strictus* are predominantly distributed in Tamil Nadu. Hence the silviculture of these two species is described.

15.13 Hollow Bamboo

Scientific Name : *Bambusa bambos*

Syn : *Bambusa arundinacea*

Comman Name : Spiny Bamboo / Hollow bamboo

Family : Poaceae / Graminae

15.13.1 Distribution

The species is one of the most important bamboos found throughout the India ascending to 1200m. It is common in Orissa, Karnataka, Tamil Nadu, Maharashtra, Kerala, and plains of Manipur. It is cultivated in U.P, West Bengal, Tamil Nadu, Karnataka, Kerala and other North-Western States. It grows well on loamy or clay loam soil. Best growth is found on alluvial soils along streams and water courses.

15.13.2 Physiognomy

It is very densely tufted bamboo with curving branches. Culms usually bright green, shining, usually 15-30 m high and 15-18 cm diameter, nodes slightly swollen, internodes usually 20-40 cm long, thick walled, branches developed from all nodes from the base upwards. Lower branches spreading and recurved spines. Spines usually in threes. Culms shealth, 15-35 x 18-30 cm. Coriaceous, with dark brown hairs, deciduous at the time of branch development. Blade up to 10 cm long, adaxial surface with dense dark brown hairs to black hairs, blade cordate, tip acute, margin ciliate.Leaves lanceolate to linear, pubescent, auriculated with silvery bristles. Inflorescence panicle often formed by whole culm, pale spikelets, lanceolate, acute, 1.25 - 2.5 cm long, 4 mm broad, sessile consisting of 1-2 empty glumes, 3-7 flowers (Plate 20).

15.13.3 Natural Regeneration

Profuse natural regeneration occurs after a flowering. Under natural conditions, the seed germinate at the commencement of the rainy season and the ground in the neighbourhood of flowered clumps becomes carpeted with green grass seedlings. Some of seedlings develop into clumps after 6 - 12 years.

15.13.4 Nursery Technique

The seeds are available in large quantities only at long intervals and do not retain their viability on storage for more than a year. Seeds are available in March - April in Central India. Plant percent is 40% and one kg of seed is expected to produce about 55000 plants.

The seed beds are prepared with well pulverized soil dug up to 30 cm. Seeds are sown in the beds in the last week of April and covered with sand. The beds are watered daily for a week then on alternate days for a month and thereafter as and when required. The seeds start germinating by about 7th day after sowing and continue for another week. Partial shade is provided to the growing seedlings. When the seedlings are about 6 months old, they are pricked out on ridges 45 cm apart and watered regularly

Plate 20. *Bambusa bambos*

by letting water along the furrows.When the seedlings have grown sufficiently in height i.e. 1.5 m to 1.75 m.,the stems are cleaned to remove side shoots and spines. The seedlings are pulled out for planting when they are more than one year old and in some cases up to 3 years. The seedlings are raised in polythene bags of size 20 cm X 10cm filled with pulverized local soil mixed with FYM.

15.13.5 Planting Seedlings

1-2 years old seedlings are planted out early in July. The seedlings are pulled out for planting out with enough care not to expose the rhizome to the sun which must remain covered with original soil. The planting distance should be according to the normal size of the clump. The spacing varies from 4m-9 m. The planting area is fenced in advance as protection to the young plants from grazing and browsing is essential. The plantation is weeded thrice in the first year and twice in the second year. Soil working around the seedlings to the radius of 30 cm and a depth of 8-10 cm is carried out.

15.13.6 Planting Rhizomes

The rhizomes are separated out carefully and transplanting the culms with 2 or 3 rhizomes attached is done.

15.13.7 Planting Offsets

In this case, one or two season old culms are given slanting cut at about 1 to 1.50 m from the ground and the rhizomes to which they are attached are dug up with roots and cut off at a length sufficient to include a bud. These offsets are planted out at 7 m x 7 m spacing. Care should be taken not to injure the buds. The planting is done in the monsoon season. The pits of 1 m^3 size are dug up and refilled with a depth of 60cm. About 3 to 4 kg of tank silt with FYM is added and a mound is formed round the culm. The rhizomes are kept in the pit at a position of 45°. The top of the culm should be cut at a slant and sealed with earth or cow dung to prevent rotting by accumulation of water. This method gives cent percent success.

15.13.8 Utilization

This bamboo is used for rafters, house posts, ladders, tent poles, shafts, mat and basket making, scaffolding, etc. Young shoots are used as vegetables and pickles. It is an important raw material for paper mills. Young twigs are eaten by elephants and buffaloes. Leaves are palatable to livestock and contain higher percentage of crude protein.

15.14. Solid Bamboo

Scientific Name : *Dendrocalamus strictus*

English Name : Solid Bamboo

Tamil Name : Kal Moongil

Family : Graminae

15.14.1 Distribution

It is the hardest, most widely distributed and important bamboo species found in India. It occurs throughout the country barring Northern parts of West Bengal, Assam and other very moist areas. It is common in Peninsular India, except very moist places. It occurs in alluvial plains and ravines, and on hill slopes to an altitude of about 1100 m.

15.14.2 Physiognomy

This bamboo has 10 to 30 culms and it is usually referred to as a clump. The culms may reach a height of 7 to 20 m and a diameter of 2.5 to 8.0 cm. The culms are almost if not entirely solid, hollowness varying from a pinhole to 2.5 cm in diameter; there is no hollowness in drier localities. The culms emerge from an underground rhizome.*D. strictus* shows sympodial type of rhizome. Sympodial rhizomes are solid usually short and thick and their lateral buds only produce new rhizomes each of which eventually produces a solitary culm. The sympodial arrangement ensures that each new culm will have its own new roots. The buds on rhizome nodes enlarge for many months in the soil and tender shoots emerge and pointed cones completely covered with imbricate sheaths. These shoots elongate rapidly attaining full height in four months which is followed by development of branches. At this stage, the culm sheaths cover the lower part of each internode and the exposed upper part is covered with waxy powder. The culm sheaths are modified leaves arranged in two ranks mainly for the protection of the growing culm. The sheath is attached at the node clasping the culm and when old, falls off leaving a scar. Width of the sheath is greater than the girth of the nodes so that the two sides of the sheath overlap. Growth of culm is rapid.Full culm maturity is completed in 3 to 4 years and new culms are produced every year with the commencement of rains (Plate 21).

15.14.3 Phenology

D. strictus exhibits both sporadic and gregarious flowering.

15.14.4 Silvicultural Characters

It is a drought resistant and frost hardy. It is also a light demander.

15.14.5 Climate and Soil

It is commonly found in drier areas and grows on well drained soils and stony soils on hill slopes.

15.14.6 Nursery Technique

Seeds are sown in raised bed. Germination commences within 11 days after sowing. Forty days after sowing, the seedlings are transferred to polybags.

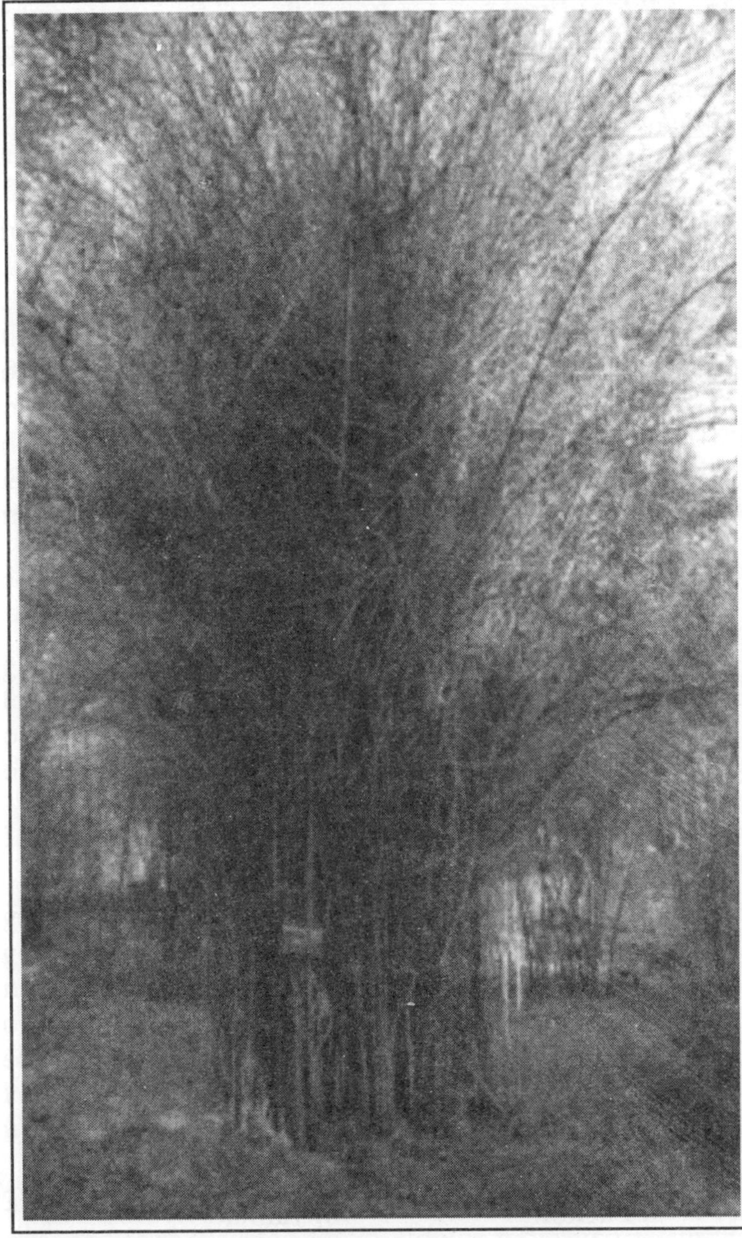

Plate 21. *Dendrocalamus strictus*

15.14.7 Planting

10 months to one year old seedlings are planted in the main field at a spacing of 5mx5m. The management of bamboo involves two main considerations viz., felling cycle and felling intensity.

15.14.8 Utilization

It is mainly used in paper industry. Its seeds are eaten by people. Leaves and young culms are used as fodder.

Questions

1. Teak is managed under

   a) Selection b) Clear felling

   c) Shelter wood d) Group selection

2. Teak is a

   a) Shade bearer b) Shade demander

   c) light demander d) None of the above

3. Casuarina was introduced first to

   a) Karwar coast b) East coast

   c) Andaman & Nicobar islands d) Himalaya

4. Loyal timber of the poor is

   a) Prosopis b) Teak

   c) Acacia d) bamboo

5. Subabul contains

   a) Lucidin b) Tannin

   c) Mimosine d) All the above

6. Teak seedlings are susceptible to

   a) Drought b) Frost

   c) Salt d) None

7. Thickest bamboo is

   a) *Dendrocalamus brandisii* b) *Dendrocalamus strictus*

   c) *Dendrocalamus giganteus* d) None of the above

8. Tiger bamboo is

   a) *Bambusa vulgaris* b) *Bambusa bambos*

   c) *Dendrocalamus hamiltonii* d) *Dendrocalamus strictus*

9. *Acacia auriculiformis* is native to

   a) Australia b) India

   c) Indonesia d) Africa

10. Tree of heaven is

    a) *Tectona grandis* b) *Tamarindus indica*

    c) *Ailanthus excelsa* d) *Bambusa bambos*

11. Indian lilac is the common name for

    a) *Azadirachta indica* b) *Tectona grandis*

    c) *Casuarina equisetifolia* d) *Acacia nilotica*

12. *Pongamia pinnata* belongs to

    a) Leguminosae b) Meliaceae

    c) Poaceae d) Verbenaceae

13. Vilayati babul is

a) *Acacia nilotica* b) *Tamarindus indica*

c) *Prosopis juliflora* d) *Acacia catechu*

14. The Queen of timbers is

a) Rosewood b) Vagai

c) Teak d) Eucalyptus

15. Beef wood is a common name for

a) *Casuarina equisetifolia* b) *Eucalyptus tereticornis*

c) *Tectona grandis* d) *Bambusa bambos*

16. Black babul is

a) *Acacia catechu* b) *Acacia auriculiformis*

c) *Acacia nilotica* d) *Acacia mearnsii*

17. Australian wattle is

a) *Acacia catechu* b) *Acacia auriculiformis*

c) *Acacia nilotica* d) *Acacia modesta*

18. Hollow bamboo is

a) *Dendrocalamus strictus* b) *Bambusa bambos*

c) *Bambusa vulgaris* d) None of the above

19. Bamboos are propagated through

a) Seeds b) Cuttings

c) Rhizomes d) All the above

20. Solid bamboo is

a) *Dendrocalamus strictus* b) *Bambusa bambos*

c) *Bambusa vulgaris* d) All the above

References

Dwivedi, A.P. 1992. Principles and Practices of Indian Silviculture. Surya publications, Dehradun.469p

Puri,S and Khosla P.K. 1993. Nursery Technology for Agroforestry-applications in Arid and Semi arid regions. Oxford and IBH Publishing company , NewDelhi. 392p.

Singh, S.P. 1995. Farming agroforestry trees. Agrotech Publishing Academy, Udhaipur. 352 p.

Tewari, D.N. 1992. A Monograph on Bamboos. International Books Distributors, Dehradun.

Troup, 1977. Silviculture of Indian trees. Vol I, II and III.

Wealth of India. 1976.Vol I and II.Published by Publication and Information Directorate, New Delhi.

—•—

